Praise for *Out of This World*

'What can your team learn from top tier leaders at NASA Mission Control? Maybe more than you think. In *Out of This World* former NASA Flight Director Paul Hill tells the true story of the game-changing transformation of Mission Control's senior leadership team. Ride along on a journey of evolution as these executives rediscover the core purpose and values that had never left their organization. Hill's candour and intensity makes this a fascinating read for every leader!'

Ken Blanchard – Co-author of *The New One Minute Manager*®¹
and *Leading at a Higher Level*²

'There is no higher-stakes environment than NASA's Mission Control. This incredible team's leadership journey – and development of precise decision-making in the face of unbelievable pressure – are inspiring. Filled with fascinating insights into spaceflight and leadership alike, every leader will find parallels to their own organization. Paul's incredible book is a must-have for anyone leading a high-performance team and an invaluable addition to any business library.'

Marshall Goldsmith – The *Thinkers50*
#1 Leadership Thinker in the World

'This is an arresting work by a former NASA Flight Director with whom I was privileged to work. Paul Hill takes the reader through NASA's legendary "Mission Control" in a way not found in any other work with which I am familiar. From its origins in aircraft flight test to the early days of the space program with Project Mercury, and on to the iconic time of Apollo, and from there to the Space Shuttle program, Paul offers a view from the inside track to both laymen and space professionals. From there, he takes you to the business world outside of NASA, and shows how the principles and values of the Mission Operations Directorate apply in a far larger arena. No leader or manager can fail to benefit from the lessons captured here.'

Michael D. Griffin, NASA Administrator, 2005–09,
and Schafer Corporation CEO

'Paul Hill has written a stunning "instructional manual" for business executives and leaders who want to learn from the best team on the planet: The men and women of NASA's Mission Control. For the first time, a leader of the Mission Operations Directorate of NASA shares the hard-won lessons of this world-famous organization and translates them into key principles and examples designed to hone a superior leadership team grounded in integrity and bedrock organizational values. Steeped in the lessons of history, rich with achievement and heart-rending loss, laser-focused on application and results, and above all a great narrative, this book, like its author, is one-of-a-kind.'

Mary Lynne Dittmar, President of the Coalition for Deep Space Exploration and former member, Human Spaceflight Committee, National Academies of Sciences, Engineering and Medicine

'This engaging book tells the story of how NASA's renowned Mission Control evolved into an extraordinary team that directed many of the world's greatest technical triumphs. Equally important is Paul Hill's cautionary tale that sustaining excellence may be more difficult than attaining it. He shares how Mission Control learned the importance of articulating, modelling and nurturing its core values of technical truth, integrity and courage to maintain exceptional performance under adverse circumstances. Leaders from every organization will benefit from these vital lessons.'

Walter E. Natemeyer, Chairman and CEO, North American Training and Development

'Mission Control is *the* leadership laboratory for those who accept the challenge and the risk of ultimate responsibility for all actions necessary for crew safety and mission success in manned spaceflight. Paul Hill is a select graduate.'

Eugene F. Kranz, former NASA Flight Director and Director of Mission Operations, 1983–94

OUT OF
THIS WORLD

How NASA created the
best team on the planet

PAUL SEAN HILL

NICHOLAS BREALEY
PUBLISHING

London • Boston

First published in Great Britain in 2018 by Nicholas Brealey Publishing
as 'Mission Control Management'
An imprint of John Murray Press
An Hachette UK Company

The paperback edition first published in Great Britain in 2019

Published in the USA as 'Leadership from Mission Control to the Boardroom'
by Atlast Press

2

A CIP catalogue record for this title is available from the British Library

ISBN 978 1 473 69610 5
Ebook ISBN 978 1 473 66848 5

Typeset in Celeste by Palimpsest Book Production Ltd, Falkirk, Stirlingshire

Printed and bound by Clays Ltd, Elcograf S.p.A.

John Murray policy is to use papers that are natural,
renewable and recyclable products and made from wood
grown in sustainable forests. The logging and manufacturing
processes are expected to conform to the environmental
regulations of the country of origin.

Nicholas Brealey Publishing
John Murray Press
Carmelite House
50 Victoria Embankment
London, EC4Y 0DZ, UK
Tel: 020 3122 6000

Nicholas Brealey Publishing
Hachette Book Group
Market Place Center, 53 State Street
Boston, MA 02109, USA
Tel: (617) 263 1834

www.nicholasbrealey.com
www.atlasexec.com

To Pam, Chelsea and Aly, who have always believed in me and
are the greatest joy in my life

About the Author

Paul Sean Hill spent 25 years in NASA's iconic Mission Control, learning and living the values he now evangelizes. As NASA's Director of Mission Operations from 2007 through 2014, Paul was responsible for all aspects of human spaceflight mission planning and training and for Mission Control. In this role, Paul led a critical leadership transformation, dramatically reduced costs and increased capability, all while still conducting highly successful missions in space.

Before this, he held a number of senior leadership positions. As a Space Shuttle and International Space Station Flight Director from 1996-2005, Paul supported 24 missions, including planning and leading space station construction in orbit and returning Shuttle to flight after the Columbia accident.

Today, Paul runs Atlas Executive Consulting and speaks frequently about universal leadership challenges. After a career leading human spaceflight operations 'from the Mission Control Room to the boardroom' he offers a candid and passionate insider's look at the leadership values and culture that have been critical for their 'impossible' wins for decades. As the executive who is credited with revolutionizing the leadership environment in Mission Control's management ranks, he shows leaders how to apply the same core ideas and values to their own challenges and boost team performance in any industry and business.

Beyond the rocket science and leadership theory, he shows how these deliberate values are the enablers in solving 'impossible' problems at all levels. Through his book *Out of This World*, leadership workshops and keynote addresses, anyone can learn how Mission

Control does it and, more importantly, how to enable a leadership environment that inherently strengthens team performance in your own organization. Paul has inspired leaders across many industries, including: Citi Group, Google, JPMorgan Chase, Shell, Texas A&M University, The Conference Board of Canada and more.

Paul lives near Houston, Texas with his family.

Contents

Acknowledgements

This book, with its leadership lessons and the stories of their discovery – and my privileged role in any of it – was only possible with the inspiration and support of a long list of people:

- The men and women of the Mission Operations Directorate (MOD), in all of its variations through the years, who shoulder the awesome responsibility, make the tough calls, shepherd our astronauts through the sky and back, and still make it look easy.
- The Apollo generation who invented the profession and the organization, led by greats such as: Chris Kraft, Gene Kranz, Glynn Lunney, Arnie Aldrich, and many more.
- The incredibly talented Mission Operations Directorate Board of Directors who saw this evolution through with me and without whom it wasn't possible: Steve Koerner, Karen Jackson, Kelly Beck, Ginger Kerrick, Jimmy Spivey, Norm Knight, Annette Hasbrook, Jean Haensley, Dan Lindner, John Sims, Brian McDonald, Jeanne Lynch, Sue Rainwater, Jim Thornton and Mark Ferring. And several notables who had moved on along the way: John McCullough, Stan Schaefer, Pete Beauregard and Mike Hess.
- G. Allen Flynt, who brought the spark that led us down the path to change everything while rediscovering who we were all along, and who also saw past my critics and helped me make my most important professional contributions.
- The thinkers and authors of the books that helped us map our recovery and refind our 'faith' as managers: Marshall

Goldsmith, Stephen M.R. Covey, John P. Kotter, Jim Collins and Jerry I. Porras.

- Milt Heflin, who in addition to his long list of accomplishments at the Johnson Space Center, was a patient mentor, was instrumental in whatever 'rounding of the sharp edges' was possible, and helped me make my way up through the management cloud many times.
- Tim Andrews and Deb Duarte, both of whom helped a recovering triple type-A Flight Director find new awareness and personal growth.
- Lee Hayward, executive coach extraordinaire, who has offered years of insights, honesty and friendship.
- Walt Natemeyer for countless leadership conversations, and for giving me the benefit of his vast expertise in situational leadership; also, for giving me the push to write this book and develop the workshops, and for tirelessly reviewing the manuscript as it materialized.
- Blair Nader, who endured me from my first day in MOD and could always be counted on to do whatever we needed from him, including reviewing the manuscript of this book and not being afraid to tell me where it needed work.
- Jack Anthony, rocket scientist and Flight Director-whisperer, spaceflight dynamics guru, unsung hero of USAF Space Command, and the kindest person you'll ever meet – who also was instrumental during the Columbia accident investigation, offered much great counsel to a NASA executive in training, and reviewed the manuscript in its roughest form.
- John M. Pretz, who showed me what real leaders look like and what it means to truly live your values in everything you do. After learning from him at the start of my career, I spent the rest trying to lead by his example.
- My Mom, who still thinks I can do anything, and who showed me years ago what it means to shoulder the weight and make your own destiny.

ACKNOWLEDGEMENTS

- My wife and daughters, who too often bore the brunt of this steely-eyed missile man's mission fixation, did not hear enough how much they meant to me, and never lost faith in me.

Preface

There is great clarity in the least forgiving, most terrifying, moments of human spaceflight. That clarity is due to the inescapable fact that some mistakes cannot be taken back, and astronauts who are counting on Mission Control may pay with their lives for any error, inattention, hesitation or less-than-perfect decision-making.

For more than 50 years, NASA's Mission Control has been known for two things: 1) perfect decision-making in extreme and unforgiving situations; and 2) producing generation after generation of steely-eyed missile men and women who continue that tradition while caring for our astronauts and the spacecraft on which they rely. A key to that legacy of brilliant performance is a particular brand of leadership. Although these leadership values are especially prevalent at the working level in Mission Control, they can also be deliberately applied to enable similar high performance in management roles in any setting.

Over the decades, as the spacecraft, rockets and missions became more complex, the men and women in the control room remained very tightly aligned to specific values, their common cause in protecting the astronauts and accomplishing the mission, and the insanely high level of individual and team performance the responsibility required. However, as essential as the leadership focus continued to be in the control room, this was not always the case in the management ranks, away from the rocket science and the mission.

Like so many large, established organizations, by the end of its first 40 years the Mission Control management team had become stove-piped at all levels from top to bottom. There was almost no semblance of real collaboration, from the common goal of protecting our astronauts to stewarding the leadership culture that is so essential to our success. As a result, by 2006 we were seeing unconscionable 'cracks in the dam' – mission-related mis-steps in management forums that would not have been tolerated in the control room, and that threatened to set the teams up for failure.

This is the story of how we became aware of the severe erosion in our management environment, and the looming risk it brought to Mission Control's perfect decision-making when it mattered most. From that sobering awareness we learned to articulate our critical leadership values, not just recognize good performance and leadership when we saw them. In the journey that followed, the management team brought the same leadership values from the Mission Control Room into our senior-management ranks and deliberately reflected them in all of our management practices.

This is how we learned to steward a culture that is aligned with, and reinforces, the awesome responsibility of our people in Mission Control when the clock is ticking. It is also how we unleashed similar stunning, Mission Control-worthy performance in our management team.

Along the way, we discovered that our problems were no different from other organizations', albeit with the potential for more spectacular and physical catastrophe than most. That discovery helped us see our problems through many different lenses, to learn from others' experience, and even to more clearly understand and articulate the principles that had always been most critical to our top performance. It also points the way for other leaders to see their problems through Mission Control's lens and apply our discoveries to their challenges.

This was our journey as a management team. It is also how to transform any leadership team in the same way Mission Control learned to transform our own, and how to develop the next generations of leaders to do the same.

This is what that looks like from the Mission Control Room to the boardroom, as experienced by the men and women who must be perfect every day in their decision-making, while protecting our astronauts. 'It ain't rocket science', as we say in Mission Control, but it can be just as difficult to grasp and apply until we learn to understand and tame the 'scary' parts.

Join us in the discovery and the journey. Learn how to leverage the Mission Control leadership values, apply them in your management ranks, and bring a culture of highly reliable decision-making to your team and your business, just as we did.

Introduction

Introducing the Leadership Journey

'Son, Mission Operations is called "Operation Head Start". It's considered a hothouse for NASA leaders. If you're coming to NASA, start there.' That advice from my dad set much of the course for the rest of my life.

After finishing graduate school and a tour as a United States Air Force satellite operations officer, in June 1990 I found myself at 28 years old in the home organization for Mission Control, NASA-Johnson Space Center's Mission Operations Directorate – 'MOD'. Although best known for Mission Control, MOD was responsible for all human spaceflight mission-planning, training and flight operations. I was hired to help plan the construction and operation of what would become the International Space Station (ISS).

The space stuff was cool, incredibly so, but I quickly learned there was much more to MOD and to becoming 'somebody' than just mastering the cool technical work – and the cool technical language. Every day in MOD, mixed in with the rocket science and space jargon, I also heard peers and managers casually using phrases such as 'steely-eyed missile man', 'tough and competent', 'conscience of human spaceflight', 'ultimate consequences', 'leadership' and . . . 'failure'.

While it sometimes sounded like they were actually speaking another language with their unique geek-speak and love of acronyms, the Mission Control veterans' straight talk and unabashed leadership focus 'had me at hello'. I felt at home right away. Now I just needed to learn the Mission Control trade, earn my place among my fellows, and maybe become a leader in this 'hothouse for NASA leaders'. No pressure.

As time went on, I noticed more and more contrast in the typical behaviours between MOD and our peer organizations in NASA's human spaceflight community, both individual and organizational. The contrast was most pronounced in the distinct moral undertone in our work, an undertone that MOD saw in righteous terms, but which others often interpreted as sanctimonious and arrogant.

Woven through much of the daily MOD business was an adamant awareness of the potential, if not likelihood, for failure. This wasn't some failure in the abstract, or something that happened to someone else. This was failure in a very personal and visceral way, one that could be truly catastrophic – blow up the rocket and kill everyone on board, perhaps even innocent bystanders if we screwed up even worse. This was failure that could never be allowed, must remain impossible, and yet had happened before. Nevertheless, that scary genie had to be kept in the bottle at all costs, and we were the ones who ultimately were responsible once the operation was underway.

Over the next six years, as I worked and learned, I was surrounded by reminders of this awesome and terrible responsibility I was now a part of. At the same time, I was struck by the reverence for leadership and problem-solving.

I worked on a variety of engineering challenges and studies as MOD participated in developing the space station design and prepared for on-orbit construction and flight operations. Throughout what seemed like endless redesigns and close calls with program cancellation, I worked on progressively more complex and comprehensive projects, and took on higher-level leadership roles within MOD.

In 1996, I was selected with four other technical leaders and managers to be one of the first International Space Station Flight Directors. This was one of those 'grab the golden ring' moments.

The public is most aware of Flight Directors from the white-vest-wearing portrayal of Gene Kranz by Ed Harris in the movie *Apollo 13*. Among other things, Flight Directors are the people who have led the teams that planned and conducted every US human space operation from Mission Control since the beginning. Most prominently, Flight Directors are in command of the flight control team and have the authority to take any action necessary to ensure the safety of the astronauts and spacecraft. I had become the fortieth in history.

I spent the next nine years as a Shuttle and ISS Flight Director, developing missions and leading operations for both spacecraft. Personally, I benefited incalculably from the 'seasoning' that comes from leading high-stakes decision-making like that, especially in the trial-by-fire that describes even normal days in Mission Control. More importantly, I also became keenly aware (and a steward) of a morality that existed in the Mission Control Room. With experience as a Flight Director, I had gone from being a willing adherent to the real-time morality and the related leadership values, to a devoted 'keeper of the faith'.

Based on that experience, *Part 1: Finding Faith – The Real-Time Morality of Spaceflight* explains how this morality evolved, how it coloured our team and leadership performance, and the distinct leadership culture that evolved over the decades. The story and the culture start in Mission Control, because this is where it matters most acutely: at the working level, in the Mission Control Room, while the clock is ticking, when the consequences of mistakes can be immediate and severe. This is when every risk, action and outcome is real, not theoretical – this is 'real-time'.

However, the values associated with this 'real-time morality' and the distinct culture are prevalent at the working level throughout the organization, not just in the Mission Control Room. Further, although these ideas and values may seem, at first blush, to matter most clearly in high stakes, technical decision-making, once understood as an integrated, philosophical foundation they are easily applied in any leadership setting. As they are applied in management roles, the wider reach of these values not only raises the performance of the organization as a whole but also reinforces the culture and performance at the working level.

In 2005, I entered the senior management and then the executive ranks in MOD. The move up and out of the Mission Control Room changed my perspective and further heightened my sensitivity to the culture, to our reputation as the 'hothouse for NASA leaders', and to a surprising reality. What I found in the senior-management forums could not have looked less like the morality and culture I had spent the previous 15 years learning and then stewarding as a near-religious experience in the Mission Control Room. The disconnect was made much more disconcerting given the moral undertone that was so correctly present in our most critical work in Mission Control. Worse, this wasn't just *any* senior-management team, this was the top tier of *MOD* management – the same MOD managers whose most fundamental responsibility was to prevent catastrophe. This was *us*.

How could this happen to *us*, the 'keepers of the faith, the conscience of human spaceflight'?

Part 2: Losing Faith – Management Clouds the Morality delves into the changes that accompany promotions into the management ranks, and the great potential those changes have for diluting our alignment to values and organizational purpose. We'll explore the difference between professed values and day-to-day management practices, and we'll learn from MOD's experience how profoundly our actual practices can drown out our values, despite our better intentions. Worse, we'll examine how the erosion worked its way down from the senior management into the organization. Not only was this chipping away at our legacy as the 'hothouse for NASA leaders', but it risked erosion in the same MOD working level that still required perfect decision-making in order to live up to our responsibilities.

In the 'sometimes it's better to be lucky than to be good' category, MOD was lucky in the selection of a new Director in 2004 who did not like what he found in our senior-leadership environment and was determined to do something about it. (This turned out to be a lucky event for me too, since it was this new Director who promoted me into MOD's senior management a year later.) That executive, G. Allen Flynt, started the dialogue and the introspection that began a genuine transformation in the MOD senior-leadership ranks.

Part 3: Refinding Faith – Transforming the Leadership Team describes this transformation, this journey. What started as a slow evolution, first latching onto a few key ideas, then catalysed more discussion, introspection and change in our management team. Through this process we brought the management team together in ways MOD had not seen in more than a decade. We learned to steward the leadership environment and culture, not simply caretake them. Ultimately, we brought the real-time morality from the Mission Control Room into our senior-management values and practices. Further, as we came to understand how essential this leadership environment was to our performance, we evolved a similarly focused process to select and develop emerging leaders into values-based leaders. Part 3 closes by streamlining the MOD experience into deliberate, step-by-step guides – a road-map to transform any leadership team and deliberately steward a culture of values-driven leadership at all levels.

This book tells our story, the Mission Control leadership transformation story. Through this journey, after almost 50 years of flying people in space, the MOD leadership team learned what it takes to manage and lead an organization as deliberately as we make the critical decisions every day in Mission Control.

The stark contrast between success and failure in spaceflight's most critical moments helps portray how essential the real-time morality is to the perfect decision-making demanded in Mission Control. But step away from the rocket science and into the management ranks, and the challenges MOD faces are not that different from any senior-management team's. Only after we made that realization were we able to turn our leadership culture around and pull off management successes commensurate with our Mission Control successes.

The morality, values, management practices and our leadership transformation, likewise, are just as applicable in any industry – not just to a bunch of rocket scientists trying desperately not to blow up the spaceships carrying their friends. The key to following us in the journey is to look past the contrasts and instead see the direct similarities, both in challenges and in opportunities to evolve and lead.

Consider this a case study from a team with a reputation for doing the impossible perfectly for decades, producing incredible leaders and stewarding a powerful leadership culture. Despite that legacy and culture, we lost our way as managers. This is how it happened, and how it can happen to other management teams. This is also how we learned to reverse the decline, bring our leadership culture back to life (and then some) and deliver highly reliable performance at all levels of the organization.

Thus, throughout the book, we'll also see the universal lessons from each step of MOD's experience through each step of our journey. This includes the 'roadmap' for evolving any leadership team, taking them through the same journey of ideas, alignment and engagement that brought the real-time morality back into Mission Control's management team.

Introducing Mission Control

Before getting to it, a little more background on Mission Control and MOD is helpful in setting the stage. An understanding of life at the working level in the Mission Control Room is essential to understanding the broader leadership culture and highly reliable organization. Similarly, a better appreciation of the origin and growth of the organization itself helps show the relevance in other management and leadership settings. It did not take long before Mission Control was not just a handful of guys flying a rocket and spacecraft. As we'll see, it evolved into a large and complex enter-prise with all of the management commitments and headaches of any large organization or company.

What is Mission Control?

Mission Control is an activity, a place, and most importantly, a team. The public persona of Mission Control, Mission Control experts and a fearless white-vested Flight Director are iconic and synonymous with technical brilliance, nerves of steel and people capable of facing any challenge and solving any problem.

Flight Control Room 1 during Gemini V, August 1965 [NASA]

Mission Control was derived from the aircraft-test community of the 1940s and 1950s. Test flights frequently used heavily instru-mented aircraft flying through the full range of flying conditions for the first time, and often at (and just beyond) the proven limits of the aircraft's capability. Test control rooms on the ground recorded the data transmitted from the aircraft and housed teams of engineers monitoring the aircraft as the pilot put it through its paces in a choreographed test plan. This ensured the aircraft remained within flyable limits, both to prevent damaging the aircraft and to protect the air crew. It also simplified the aircraft design by reducing the need to record aircraft system and test data on board, ensured the data survived any accident or failure during testing, and enabled a much wider array of experts to evaluate data and lend their judgement during test flights than the air crew alone.

The same needs led to a Mission Control from the start of the US ventures in human spaceflight. As rockets and spacecraft grew more complex, they quickly outpaced the flight computers' ability to monitor all systems, perform high-volume calculations and present the necessary data to the flight crew. Bridging that gap, Mission Control took on a greater responsibility for flight control, or direct control of the vehicle and rocket, as well as guidance to the flight crew based on the greater insight afforded to the Mission Control Center (MCC) with its powerful computing capability and large team of experts.

Starting with Mercury, Gemini and Apollo, the full range of human mission-phases was invented, tested and leveraged to do more in stepping-stone fashion. From 1961 to 1969, missions evolved from sub-orbital to Earth orbiting. In that same time frame, NASA went on to rendezvous, docking, spacewalks, lunar orbit and lunar surface operations. By 1973, this expanded to months-long missions in low Earth orbit on the Skylab orbiting outpost, and in 1975 the first international mission, Apollo-Soyuz, between Cold War enemies.

In 1981, NASA began 30 years of Space Shuttle operations. Shuttles deployed satellites, conducted myriad scientific experiments, perfected spacewalking and robotics, and returned to Earth with thousands of pounds of salvaged satellites, science instruments and other payloads.

Those efforts ultimately led to an unprecedented global peacetime effort to build the International Space Station. Construction began in 1998, leading to permanent manning two years later and culminating in almost 1 million pounds of spaceship assembled in low Earth Orbit while travelling 17,500 miles per hour.

With the increased mission complexity came increasingly harsh flight environments and correspondingly more complex spacecraft.

Our astronauts travelled further, faster and with more equipment to do more demanding work. This led to increasingly complex flight plans, astronaut and flight controller training, and flight operations, along with a wide array of computers, networks and training systems required behind the scenes. Like the complexity and the risks, the MCC evolved and grew significantly, as did MOD's responsibilities and organization.

In 1958, Project Mercury Mission Control was built as a small complex near the launch site at Cape Canaveral Air Station, Florida. This site was expanded in 1962 and 1963 for the first three Gemini Project flights.

Mercury Control, Cape Canaveral, June 1963 [NASA]

In preparation for a much heavier mission-oversight load in the Apollo Program, a new MCC was built at the new Manned Spacecraft Center in Houston, Texas (renamed the Lyndon B. Johnson Space Center, JSC, in 1973). Every NASA human space mission since Gemini IV in 1965 has been controlled from the JSC MCC. In its JSC home, the MCC grew progressively into a 5-storey computing and communications hub, driven by mainframe computers, routing data, voice communications and video through miles of cables, and housing hundreds of flight controllers, support engineers and managers.

Mission Control under construction at JSC in 1963 and Flight Control Room-1 in 1965 [NASA]

Mission Control and the ISS flight control team in Flight Control Room-1, 2009 [NASA]

Throughout these five decades, rockets and spaceships became increasingly complex, as did our missions. To keep pace, not only did the MCC become more complex, but the full MOD portfolio also required more effort, investment and people to do the job without lowering the bar in the rocket science. Along the way, Mission Control grew into a very large organization, both in infrastructure and in people.

What began as Project Mercury's Flight Operations Office, with 20 scientists and engineers led by Walt Williams at Cape Canaveral, grew into JSC's Flight Operations Directorate during

Apollo. It expanded again into the Mission Operations Directorate in 1983, peaking at just over 5,000 employees in the early 1990s while flying Shuttle missions and preparing for space station operations.

Apollo Flight Director Glynn Lunney 'on console' during Apollo 13 [NASA]

MOD's leadership culture was born first in the personalities of some key leaders who invented the profession and the organization, including Walt Williams, Chris Kraft (the first Flight Director and future JSC Center Director), Bill Tyndall, Gene Kranz and Glynn Lunney, just to name a few. They were tough-talking, no-nonsense quick studies and strong engineers who took on the creation of a flight operations organization that would become Mission Control and then MOD. More importantly, these men embraced the daunting responsibility they held for the astronauts' lives. They were also committed to the original human spaceflight mission with near-religious fervour in the name of national pride and a fallen President, John F. Kennedy.

Like Mission Control and MOD, the rest of NASA grew and changed dramatically in those same five decades. Space programs came and went.

Management fads went in and out of vogue. As a government

agency, NASA periodically lost momentum and strategic focus as political support and direction from above ebbed and flowed.

With all of this change and growth over the years, what didn't change? Mission Control maintained a keen focus on exceptional performance when the clock is ticking and our astronauts are in harm's way.

The key to Mission Control's success and legacy was never the infrastructure. Although critical to doing the job, the MCC, computers, communications networks etc have always just been tools of the trade. The persistent focus on performance and leadership in the Mission Control Room has always been our greatest strength and our defining cultural characteristic.

We learned much of this early in the business in the school of hard knocks. As the Mission Control teams progressed through increasingly difficult missions, struggled with technical challenges, learned (invented) the ropes and experienced and rebounded from catastrophe, the original teams codified 'normal' ways of doing business and Mission Control values.

As we'll see, we 'found our religion' or 'faith' in real-time and set the cultural foundation for the organization. Over the years, we grew significantly into a much larger and more complex enterprise and 'lost our faith' along the way in our management ranks. However, after more than 40 years, we learned how to articulate not just the morality that had always been so critical in Mission Control but also the foundational values for the larger enterprise – the organization. We 'refound our faith' as leaders. From there, we quickly learned to apply it in our management practices, and most importantly, to deliberately steward the values and culture.

This is how we brought our core leadership values and highly reliable performance from the Mission Control Room to the boardroom. It's also how anyone can do it, benefiting from our successes and failures.

It ain't rocket science . . .

PART 1

....

Finding Faith

The Real-Time Morality of Spaceflight

All too often, successful missions and sometimes even crew and spacecraft survival are delivered on the strength of the operations team. That strength is found in the team's expertise and preparation as well as on proven leadership. This is the kind of leadership environment where it goes without saying that everyone on the team will deliberately and consistently do the right thing for the right reason; will work to be as good as they can in whatever role they serve; and are willing to step up and make the call.

Paul Sean Hill, Director of Mission Operations, 2010

PART 2

Finding Faith

CHAPTER 1

Clear Alignment to Purpose

We are the last line of defense for our crews.

Eugene F. Kranz, Flight Director and
Flight Control Branch Chief, 1965[3]

Start-up and Organic Evolution

What is our purpose? Ask most people that question about their organization, whether in the start-up phase or as a mature, ongoing enterprise, and you're likely to get a list of jobs and products.

Mission Control is no different. In fact, setting out to invent Mission Control in 1958, the leaders first had to grapple with just defining the job. They were faced with all kinds of basic questions, as our first Flight Director and Apollo legend, Chris Kraft, said in a lecture at MIT in 2005[4]:

How many times around the Earth do you think we would like to go or need to go on the first flight? And what do you think would determine that?

What is a real-time decision? Where are you going to make a decision? Do you need some central facility (which invented Mission Control)?

And if you're going to make decisions in a central location, then you've got to have some means of getting that data back to them, of massaging that data, letting people know outside the limits of that control facility what is going on so they can interrelate with each other.

Nobody had ever done that before.

If this system is failing, what are the measurements that we're going to have there? And, if it is failing, and it isn't operating at the right temperature or the right pressure, and it is off nominal, what will the system do? And how do we measure that on the ground? How do we detect it?

Starting with a litany of questions, they took an inventory of the jobs that had to be done in real-time – that time when the clock is ticking, the rocket is burning, the spacecraft is in the sky and the astronauts are in harm's way. Based on those jobs, they defined the computers and communications networks they would need, and the myriad other mission planning and training tasks required before the first astronaut strapped himself onto the first live rocket.

That's how they knew which things to develop first, how many people to hire etc. All of which is essential management stuff. What they didn't do was spend a lot of time philosophizing about some underlying core purpose. That would come over time, although even 45 years later MOD leaders would typically still answer in terms of jobs and products.

Unsurprisingly, through the ensuing decades there was a considerable list of jobs just on the flight control team while flying the spaceship and looking after the astronauts in real-time. In normal or 'routine' human spaceflight, Mission Control's typical tasks include:

- Updating computer sequences that control the rocket and spacecraft systems.

- Comparing the spacecraft's guidance and navigation calculations against separate calculations in the MCC and updating rocket engine controllers.
- Balancing electrical power generation and usage across multiple power sources, managing thermal control systems to ensure hot areas are cooled and cold areas are heated.
- Managing the use of consumables such as hydrogen, propellants, coolant and food, the use and production of others such as oxygen and water, and the storage and disposal of things such as human waste.
- Monitoring atmosphere composition, adjusting the introduction of air, oxygen and nitrogen into the cabin, and managing the systems that remove carbon dioxide and other contaminants from the air.
- Monitoring satellites and experiments carried on board, from initial activation to full service, and routing the science data to the ground.
- Choreographing robotic operations and spacewalks, monitoring spacesuits and the systems required for the astronauts to step into space as well as the tools they use while outside.
- Monitoring the astronauts' health and guiding them through any treatments.
- Monitoring space radiation, which affects both crew health and spacecraft electronics, and advising the crew on actions required to minimize exposure.
- Monitoring weather at all potential landing sites, targeting landing in areas that are within wind, precipitation, visibility and sea-state limits, and coordinating with ground recovery forces to ensure the crew is able to safely exit the vehicle.

But when things don't go so normally, Mission Control adds to their focus:
- Replanning the day, the next day and the rest of the mission in response to system failure, operations running long or changes in mission objectives.

- Intervening by sending a command to power off a failing piece of equipment, switch to a back-up, reset computers and automatic sequences, command an abort to separate the crew-carrying spacecraft from a failing booster, confirm false fire alarms, manage the systems required to cleanse the air after a fire in the cabin, and much, much more.
- Developing maintenance and repair procedures to restore systems in flight.
- Ideally, before any of these occur, recognizing signs of equipment behaving in a way that could lead to failure and taking action to prevent the failure from ever happening.

These tasks are spread out across a flight control team of individual flight controllers in each position in Mission Control. Each flight controller is responsible for a specific system or technical discipline, for example, the electrical, propulsion, robotics, spacewalks etc.

Progressing from mission to mission, we learn from our successes and failures. As we solve problems and learn what works, we reinforce those decisions and behaviours in order to contribute to future successes. In less successful outcomes, we take note of those 'that could have gone better' moments. Over time, we expand the Mission Control community awareness of lessons learned and best practices for solving specific problems when faced with them in real-time – best ways to focus on critical information, evaluate and manage risks, and communicate it while conducting a rapidly changing operation. We evolve our standards for being a flight controller and part of the Mission Control team.

Again, this all sounds very cool, but it's still just a list of 'What we do'. As important as that list is, and as important as evolving our best practices and operating standards is, they are not enough to justify the moral undertone or morality in the Mission Control Room. That morality emerged when Mission Control started flying, rather than just crunching numbers and planning to fly.

▶ Each of the specific jobs that comprise our role is important and must be done correctly.

▶ No matter how critical or difficult they are, the jobs themselves do not define a morality – a framework by which performance is judged.

Experience Leads to the Foundations of Mission Operations

Beyond the details of the job, the experience of real-time operations has a way of focusing our attention on what is most important. It is one thing to sit around a conference room talking about some dangerous operation in the future. It is another thing to find yourself 'on console', trying to keep up with the spaceship and stay ahead of events as they unfold, all while a crew of astronauts is living with the results of your efforts and hurtling through sky and space. A defining result as Mission Control gained real-time experience was a clear understanding of our core purpose and what it took for us to deliver.

The earliest hints of there being something more to the job came unexpectedly from the morale problems that accompanied the failures in the early days of Project Mercury. Leading up to Alan Shepard's historic Mercury Redstone 3 launch on 5 May 1961, only 8 of the 17 preceding uncrewed Mercury missions had been successful. In the weeks before launch, the Mission Control team was feeling the pressure from the daily criticism in the press. More acutely, as Shepard's launch approached, was the growing reality of how dangerous this mission was for him.

In addition to their technical work, Flight Director Chris Kraft paid close attention to the team's morale, specifically ensuring they remained focused on the job at hand and didn't let any stress-induced mistakes creep into their mission preparation. Mission Control had to be ready during the mission. When the clock was ticking, it would be on their shoulders to beat the previous failure rate . . . to keep Shepard alive.

That awareness stuck with the Mission Control teams as they progressed and learned through the six manned Mercury missions. Four years later, during final preparations for Gemini 3 in March 1965, a different management challenge presented itself, also not technical. A conflict flared up between a flight controller and an astronaut at a remote site in Carnarvon, Australia. Flight Control Branch Chief Gene Kranz had put the flight controller in charge at that site during the mission. Coincidentally, Chief Astronaut Deke Slayton had told a new astronaut he was in charge.

Thus began a several-day squabble that almost turned into civil war between the flight controller and the astronaut for the right to be the boss during the mission. In addition to spilling over to their managers, Chris Kraft and Deke Slayton, the dispute also distracted the flight control team at the L-1 day launch briefing of only the seventh human mission in US history.

Fortunately, Gemini 3 went on to fly without a hitch. However, after the mission, Gene Kranz admonished the full flight control team[5]:

Discipline. If you remember only one thing from this debriefing, I want you to remember one word . . . discipline! Controllers require judgment, cool heads, and they must lead their team. Leadership, judgment, and a cool head were not evident at Carnarvon. I understand how conflicts can raise tempers, but I also expect control.

We are the last line of defense for our crews. We let a small incident escalate into a major flap that involved the entire operation. It was a distraction that we should have put behind us. We lost track of our objective.

Our mission will always come first. Nothing must get between our mission and us . . . nothing! Discipline is the mark of a great controller.

The fallout from that incident created a long-lasting rift for the first time between the astronaut office and the flight controllers. To restore the peace after the mission, the flight controller was transferred out of Mission Control to another NASA center.

Two years and nine Gemini missions followed at a dizzying rate as NASA learned how to perform the more complicated operations required to fly to the Moon. On 27 January 1967, while preparing for the Apollo 1 launch scheduled for less than a month later, NASA's worst fears were realized. An oxygen-driven flash fire erupted in the capsule during a test on the launch pad. Mission Control could only listen while astronauts Gus Grissom, Ed White and Roger Chaffee died in under 30 seconds.

This test did not have Mission Control in the role of leading or even officially participating. They were only monitoring as part of their pre-flight preparations, so it was not a 'real-time' experience for the MCC. But it was definitely real-time for the crew, the three men who, on this day, Mission Control and NASA were not able to protect.

The flight control team was stunned by the loss. Not least among them was Flight Director Gene Kranz. He immediately realized that Mission Control could not ignore their share of the responsibility for failing the crew. In his words after the debriefing three days later[6]:

Spaceflight will never tolerate carelessness, incapacity, and neglect. Somewhere, somehow, we screwed up. It could have been in design, build, or test. Whatever it was, we should have caught it.

We were too gung-ho about the schedule, and we locked out all of the problems we saw each day in our work. Every element of the program was in trouble, and so were we. The simulators were not working, Mission Control was behind in virtually every area, and the flight and test procedures changed daily. Nothing we did had any shelf life. Not one of us stood up and said, 'Dammit, stop!'

I don't know what Thompson's committee [Dr. Floyd L. Thompson, Chairman of NASA's Apollo 1 review board] will find as the cause, but I know what I find. We are the cause! We were not ready! We did not do our job. We were rolling the dice, hoping that things would come together by launch day, when in our hearts we knew it would take a miracle. We were pushing the schedule and betting that the Cape would slip before we did.

From this day forward, Flight Control will be known by two words: Tough and Competent.

Tough means we are forever accountable for what we do or what we fail to do. We will never again compromise our responsibilities. Every time we walk into Mission Control we will know what we stand for.

Competent means we will never take anything for granted. We will never be found short in our knowledge and in our skills. Mission Control will be perfect.

When you leave this meeting today, you will go to your office, and the first thing you will do there is to write 'Tough and Competent' on your blackboards. It will never be erased. Each day, when you enter the room, these words will remind you of the price paid by Grissom, White, and Chaffee. These words are the price of admission to the ranks of Mission Control.

Twenty-three years later, in 1990, when I started in MOD, 'tough and competent' was still shorthand for sucking it up, doing the hard jobs and making the call, unpopular or not. More importantly came the realization in 1967 that Mission Control's influence had to extend beyond the mission plans, procedures and operations. With spacecraft, rockets and missions becoming so much more complicated, it was imperative for flight controllers, engineers outside of Mission Control and managers to voice their concerns at all phases of development and mission preparations, just as it is for flight controllers during real-time.

After the herculean effort of the Mercury, Gemini and Apollo Programs, the flight rate and development pace slowed to a relative crawl. NASA flew only three flights to Skylab between May 1973 and February 1974, followed by the joint US–Soviet Apollo-Soyuz docking mission in July 1975. The Space Shuttle's first flight was not until a distant April 1981. MOD managers worried that the long time without flying and the hand-off from many of the Apollo veterans to a new generation in MOD could lead to a loss of the focus and values upon which success depended. In 1979, Gene Kranz wrote a memorandum to the operations community encouraging them all to step up to the challenge and responsibility of the first Shuttle flight, invoking words from Mercury, Gemini and Apollo: *morale, discipline, tough and competent.*

Over the next year, Flight Director and Flight Control Division Chief Pete Frank dwelled on those words and the concerns that led to them. He sought to articulate a more complete creed, not incrementally after a series of bad experiences and failures, but with the foresight and thoroughness demanded of Mission Control. Pete wrote in a 1980 memorandum[7]:

> We have been working in a 'pre-flight mode' for so long that I think many of us have forgotten what a real spaceflight operation requires. In the course of what seems to be endless meetings, simulations, schematic preparation, etc., it seems as though the actual flight will always be in the future. As a result, we tend to get more casual about our responsibilities and tolerate performance in ourselves and others that we would never accept in a 'flight mode' environment. I think it's time to shake off that attitude and remind ourselves what is going to be expected of us in the near future.

> . . . I am distributing a copy of 'The Foundations of Flight Control', which is an attempt to summarize the ideas Gene discussed. It develops the theme into a daily reminder of who we are, what we

stand for collectively as a division, and what we should strive for as individuals.

And so was born in 1980 The Foundations of Flight Control, which have since evolved into 'The Foundations of Mission Operations'. The original words, all of which are unchanged, are from Pete Frank, with much of the theme and passion clearly influenced by Gene.

One addition to The Foundations followed the Columbia accident and the loss of seven more astronauts on 1 February 2003. The Columbia Accident Investigation Board's final report cited, among other criticisms and organizational contributors to that accident, the 'reliance on past success as a substitute for sound engineering practices'.

Two years later, just before the successful Shuttle return-to-flight, Chief Flight Director Jeff Hanley added *vigilance* to the Foundations. Republished almost 44 years after Alan Shepard's flight, this addition was the reminder that success and experience in a dangerous endeavour do not render it safe or even less dangerous. As Jeff wrote to the Astronaut Office when the revised Foundations were distributed[8]:

The unique challenge of the Space Shuttle and ISS programs has been in the long timescales through which we must maintain that vigilance. To embrace that commitment and keep it strong, we have adopted 'vigilance' as a key fundamental quality essential to our professional excellence. Our 'vigilance' will be focused always on remaining attentive to the dangers of spaceflight, while never accepting success as a substitute for rigor in everything we do.

The Foundations of Mission Operations

1. To instil within ourselves these qualities essential to professional excellence

Discipline Being able to follow as well as to lead, knowing that we must master ourselves before we can master our task.

Competence There being no substitute for total preparation and complete dedication, for space will not tolerate the careless or indifferent.

Confidence Believing in ourselves as well as others, knowing that we must master fear and hesitation before we can succeed.

Responsibility Realizing that it cannot be shifted to others, for it belongs to each of us; we must answer for what we do or fail to do.

Toughness Taking a stand when we must; to try again, and again, even if it means following a more difficult path.

Teamwork Respecting and utilizing the abilities of others, realizing that we work toward a common goal, for success depends upon the efforts of all.

Vigilance   Always attentive to the dangers of spaceflight; never accepting success as a substitute for rigor in everything we do.

2. To always be aware that suddenly and unexpectedly we may find ourselves in a role where our performance has ultimate consequences.

3. To recognize that the greatest error is not to have tried and failed, but that in the trying we do not give it our best effort.

The Foundations resonated with the Mission Control team in 1980, just as they do today. They can be found posted in the hallways, offices, workbooks and websites. They are routinely referenced and paraphrased in casual conversation and in tough technical deliberation.

The Foundations of Mission Operations are ubiquitous reminders in Mission Control (and MOD-wide) of our responsibility to protect the crew and return them safely to the ground at all costs. They are intended to never let us forget that the ultimate consequences for which we may find ourselves playing a role are paid by the astronauts if we do not give our best effort. They also remind us of the essential qualities that are both necessary for success and that we can rely upon in those perilous, danger-filled moments in real-time when, as Gene Kranz said after Gemini 3, 'We are the last line of defense for our crews.'

▸ Learning from experience – triumphs, mistakes and failures – greatly clarifies the focus that is required to be successful and our role in making it happen.
▸ By articulating that understanding, The Foundations of Mission Operations gave first voice to Mission Control's moral underpinning.

Our Purpose Is More Than What We Do

Officially and formally, MOD is responsible for mission planning, training and flight operations for NASA's human spaceflight programs. Flight operations – Mission Control – is our real-time and fundamental role. To go beyond the individual jobs and define a morality, we need something more philosophical and existential, something that answers why it matters, and why it matters to do it well. The Foundations of Mission Operations are a deliberate reminder of the moral imperative in everything we do because of the direct effect every decision and action could have on our crews' lives and on accomplishing increasingly complex missions.

In Mission Control, we are reminded every day that success or failure in our most basic purpose is at stake in everything we do in this incredibly dangerous business, and people's lives hang in the balance with any failure. Screw up a decision, stop paying attention and miss some equipment failure, ignore some warning sign because it isn't convenient in today's plan . . . and damn it, we are forced to abort a mission, having now put the crew into harm's way for nothing.

Or damage the spacecraft and put the crew at even greater risk.
Or we kill the crew.
Failure.

Worst of all, and inexcusably, we could fail in situations that are otherwise within our grasp if we remain focused and aligned in our decision-making.

Beyond the rocket science and engineering, after debates in conference rooms, when the mission is underway, we distil this down to:

Mission Control's Core Purpose: Protect the crew and return them safely to the ground at all costs. Ideally, in doing this, prevent damage to the spacecraft, and achieve as many of the mission objectives as possible, in that order.

Everyone on the team is firmly aligned to this purpose. As each of us gains experience, it is burned into our psyche. Your specific job and level of seniority are secondary to that alignment and sense of ownership. Each person on the team is responsible to Mission Control's core purpose, and their every decision is judged accordingly, as is the entire team's.

....

Full Alignment and Ownership

As an example of how powerful this alignment is, consider a training run (simulation) in Mission Control one day during my first year as a Shuttle Flight Director. Simulations are typically tough – much tougher than a normal day of flying in space. The training team throws all kinds of failures at the flight control team, sometimes with only a very narrow way through the woods to success, and it's up to the team to find it . . . protecting the crew, the spacecraft, and then the mission. This was definitely one of those days.

We had navigation system problems. We were dealing with electrical, computer and propulsion system failures. We were racing to keep the Shuttle in flying condition, while at the same time executing a series of rocket engine burns necessary to rendezvous and dock with the Russian Mir space station at more than 17,500 miles per hour. All of which was pretty much just another day in training for human spaceflight.

We had already made the largest burns of the series and put the Shuttle on a collision course with Mir. In doing so, we had used enough propellant that we were now in an 'all or nothing' situation. If something happened that forced us to abort the rendezvous, we couldn't try again. We'd have to land the Shuttle without delivering our cargo. At the same time, the astronauts were now preparing to finish the rendezvous with a final set of burns and some close-in stick-flying.

As luck and the training team would have it, just after we had crossed that line of no return, a smoke detector went off, signalling a potential fire in one of the Shuttle avionics bays. *Potentially* not good. Besides the problems with flame and smoke in the crew cabin, a fire in one of the avionics bays would damage critical systems we needed to maintain control of Shuttle. OK – all of that could potentially be really, really bad.

As I discussed the smoke detector with the flight controller responsible for that system, EECOM, Shuttle was still closing on Mir, and we had business to take care of to help the crew get there. The discussion between me (Flight), EECOM, the Rendezvous officer (Rndz) who was responsible for trajectory guidance in this phase of the approach, and the Propulsion Officer (Prop) went like this, most of it on our recorded voice loops, as we are trained to do:

Flight: EECOM, how many active indications of smoke do we have on board?

EECOM: Just one, Flight, in the Av Bay ventilation duct.

Flight: And the crew reported they didn't smell any smoke when the alarm sounded . . .

Rndz: Flight, we see the crew targeting the next burn in the manual phase. It looks good, and we're go to burn the on-board solution.

Flight: Capcom, let the Commander know they're go for the burn with the on-board.

Capcom: Endeavour, Houston. We see your on-board solution. You are go for the burn.

Endeavour: Copy, Houston.

Flight: Rndz, any concerns with radar data we're taking?

Rndz: No, Flight. Looks fine now.

Flight: EECOM, what other options do we have for secondary cues for smoke in this Av Bay?

Prop: [shouting over the console to the room] Does anyone else care that we are still flying this rendezvous with a fire on board?!

Flight: Sit down, Prop. I'll tell you when we have a fire.

That cracked me up at the time. Still does, every time I tell the story, although at the time it pissed off the Propulsion Officer. She didn't care that smoke detection was outside of her area of responsibility. She was reminding me that *our* first priority was protecting the astronauts, not flying a burning wreck to Mir at all costs.

Of course, we weren't ignoring a raging inferno on board. The EECOM and I were discussing the other smoke detectors we should see going off if we had some piece of equipment smoking or on fire in that avionics bay. Also, the air in this avionics bay was continuously circulated into the cabin, where the astronauts were not smelling smoke or smouldering insulation. With no other smoke detectors sounding off, the astronauts not smelling anything, and the equipment operating normally in that avionics bay, there was an almost certainty that we did not have a fire on board. The details were understandably not as familiar to Prop as they were to me and the real expert, EECOM.

With that in mind, we couldn't abort the rendezvous for what was likely a false alarm from a failing smoke detector or caused by one of our previous computer failures. Do that, and not only would the crew be no safer than they already were, but we'd definitely blow the mission.

In the meantime we had some of the astronauts already at work opening the avionics bay (not an easy or short procedure) so they could take a look, while the others continued flying. If necessary, there wasn't a single action we needed to put out a fire that we couldn't do while Shuttle continued the approach to Mir. If, in the end, we did have a fire, we'd simply put it out, stop the approach, fly away from Mir and begin planning a landing, probably the next day.

....

The moral to the story is all about the alignment to core purpose. The Propulsion officer is responsible for the propulsion system. EECOM is responsible for smoke and fire detection. As the Flight Director, I was officially responsible for everything. However, Prop, like everyone on the Mission Control team, is also responsible for everything. That includes poking the Flight Director – the boss – to ensure I hadn't lost my mind and stopped making decisions according to our most basic value, our core purpose, our moral imperative – and hers and the entire team's: Protect the crew and the spacecraft first.

I can't remember if we completed that rendezvous or ultimately had to abort. I also couldn't say who a single other member of that day's flight control team was. However, I vividly remember the exchange with Propulsion Officer, Cathy Koerner, who was unafraid to call me out while we were flying – the fact that we were already doing the right thing notwithstanding. It was a strong reminder for the rest of the team.

That's alignment to core purpose, and it is played out every day just like that in Mission Control. This particular example was just one of many like it that got Cathy Koerner selected to be a Flight Director herself a few years later.

Clear alignment to Mission Control's core purpose is where the real-time morality starts. Protect the astronauts, protect the spacecraft, and then accomplish the mission. How we judge ourselves and each other is based on doing this every time, in every decision and action, individually and as a team.

▶ Full alignment to core purpose is where team morality starts.
▶ How we judge ourselves and each other is based first on the extent to which the alignment to the core purpose is reflected in every decision and action, individually and as a team.

Why We Lead

In learning to articulate The Foundations of Mission Operations and reinforce alignment to Mission Control's core purpose, we have always simply been trying to understand the answers to some basic questions:

- Beyond the specific jobs, what are we really trying to accomplish?
- What is it we must excel at?
- In order to achieve anything, what is it we cannot fail at?
- What makes us think we're delivering?
- How do we keep delivering?

In other words, what is our real goal? What is our leadership intended to provide? How do we ensure that's what we're doing? Why do we lead?

It is not difficult to see how Mission Control's core purpose has the power of a moral imperative. The compelling image of controlling the rocket, flying the spacecraft and protecting our astronauts while accomplishing missions in space is a strong, almost organic, aligning catalyst as we bring in new generations to take the reins.

Let's see . . . manage the rocket, spewing millions of pounds of thrust, hurling our friends into the sky to return later in a ball of fire, all with such narrow margins between success and failure that perfect team performance is required. Critique every nuance and decision with that in mind, and align all judgement towards protecting the astronauts at all costs? Check. No convincing required. Complete alignment in that core purpose is essential.

When I explain the Mission Control leadership culture in speeches outside NASA, group after group consistently sees how crucial this alignment is in *our* rocket science, where small errors can result in enormous and terrible failure. However, some people struggle to find the relevance in their own work, which in many cases is much less scary and unforgiving than rocket science. The dilemma I've had posed to me many times is, 'Hey, I can see how this matters to Mission Control. But we don't fly rockets and spaceships. We're not going to kill the astronauts or anyone else if we make mistakes. How does this matter to us?'

Look past the smoke and fire, astronauts etc that define Mission Control's specific occupation. Why does anyone lead?

Your business or organization has some goal, some product or service it is intended to provide for some value. That value can be measured in many different ways. It may be in the marketplace and look like profit. Or, like NASA's, could be responsible use of taxpayers' investment to accomplish some national objective. This would be the base product or service, without which your organization would not exist. After that, avoid mistakes that are so severe the enterprise cannot ever afford to make them.

Many lines of work other than spaceflight involve high energies and risk to people and infrastructure (oil and gas exploration, air traffic control, and many more). The comparison to Mission Control doesn't apply only to them, though. More fundamentally, *we all* seek to prevent actions that erode the trust of customers and stakeholders – things that not only cost the organization support but could also put it out of business. In the purest business sense, don't let something slip by that bankrupts the company or in some way erases all previous accomplishments. Get *your* rocket science right. Don't blow up *your* rocket, however figurative. Avoid catastrophe.

Then learn from your experiences to catalyse strategic innovation – get better at *your* rocket science, and use that improvement to extend your capability. 'Better' could mean 'more efficient', 'higher quality', 'faster production' etc. It also means leveraging the insights gained through different experiences to solve problems previously considered unsolvable. In that way, learn enough to extrapolate to new products and new markets.

Finally, align the organization to these same objectives: the mission, avoiding catastrophe and catalysing strategic innovation. For Mission Control's level of effectiveness, this means buy-in and ownership at all levels, not just in the formal leadership ranks.

Why We Lead	In Mission Control	Anywhere
Accomplish a mission, add value	Accomplish all mission objectives	Deliver a product, provide a service – satisfy the customer
Prevent catastrophe	Protect the astronauts, the spacecraft and the public (Don't blow up the rocket or crash the spaceship!)	Don't bankrupt the company or damage the company's reputation with customers and stakeholders
Catalyse strategic innovation	Learn from each experience to improve, solve problems previously considered impossible, and enable new missions	Increase efficiency, learn from experience to enable new products and new markets
Continuously reinforce alignment at all levels to these same objectives		

In Mission Control, it is a great leadership luxury to have such a clear and compelling purpose with such a readily discernible moral component. That clarity simply makes it easier to see the imperative for us to align to our core purpose. It does not weaken the value in other venues. Some may just have to work harder to answer the question, 'How does this matter to us?'

CHAPTER 2

Demonstrated Behaviours that Reinforce Alignment

In the majority of men, success or failure is caused more by mental attitude than by mental capacity. The willingness and ability to strive, and to do, are best judged by what we see of men in action.

General George C. Marshall, Secretary of Defense, 1950[9]

So What's So Hard About That?

In Mission Control's case, much of the work is still *actual* rocket science.

In its infancy in 1961, riding into the sky on a missile was certainly not a safe or easy endeavour. More than 50 years later, flying in space remains exceptionally dangerous, even during 'routine' flight with everything operating as designed.

Yes, the current generation starts with an experience base the Apollo generation had to first gain. However, consider the relative significance of our total experience in flying people in space compared to commercial air travel as an example of scale. By 2016, a little more than 300 human missions have been launched into space worldwide *in history*, and only half of these were from NASA and Mission Control. Contrast that to more than 60,000 commercial airplane flights in the United States, and 100,000 worldwide, *every day.* As experienced as today's Mission Control is compared to the 1960s, we are still managing enormous risks in a comparatively unexplored environment.

It is also true that technology has improved in most technical areas, especially in computers, communications and navigation systems. Like our accumulated experience, the technological advances have contributed to greater reliability and confidence in operational decision-making.

Nevertheless, everything that reaches the lowest and most 'easily' achieved Earth orbit possible must still travel almost 17,500 miles per hour – 25 times the speed of sound. The energy required to do that is massive. Even the most modern rocket engines generate huge amounts of thrust as long columns of fire in order to produce that much energy. Their *millions* of pounds of fire must be precisely controlled both to make it to orbit and to prevent the rocket from blowing up along the way. Even just this step is barely within human capability to do safely enough for human missions.

For a better idea of the level of difficulty, let's put some measurements against that, using the Space Shuttle as an example:

- To reach orbit, Shuttle left the pad with 7 million pounds of fire blasting out of the back end of a 4.5 million-pound launch vehicle.
- Two minutes later, the spent solid rocket boosters dropped off at 4.5 times the speed of sound, and the fire leaving the vehicle dropped to 'only' 1 million pounds.
- As the main engines continued burning, they turned liquid hydrogen and oxygen into 6,000 degree F gas at pressures as high as 6,000 psi, hot enough not just to melt iron but to boil it, and more than 400 times the air pressure on the surface of the Earth.
- Eight minutes later, the main engines cut off, and the 200,000+ pound spacecraft is more than 100 miles above the Earth, hurtling through the sky fast enough to cover more than 85 football fields every second.

- And it's in space . . . no air, extreme cold and heat, cosmic and solar radiation, space junk and micrometeoroids whizzing by, and more.
- At the end of the mission, the Shuttle slowed down by less than 200 miles per hour, enough to drop into the upper atmosphere, where a fireball forms around the spacecraft, further slowing it to less than 9,000 miles per hour in under 20 minutes.
- This fireball heats up some parts of the Shuttle's outer surface to temperatures just below 3,000 degrees F, well above the melting point of any of the metals in the spacecraft structure.
- After that, the 200,000+ pound, engineless Shuttle glides more than a quarter of the way around the world for a one-shot landing.

Incidentally, compared to a launch-and-entry vehicle such as Shuttle, flying along inside a comparatively sedate orbiting vehicle such as the International Space Station is still a scary business. We may sometimes joke about 'only boring holes in the sky' as ISS endlessly orbits the Earth, but it is almost a million-pound spacecraft, assembled largely by hand in orbit. It lives out its life continuously exposed to the space environment, and now we have astronauts living there for 6 to 12 months at a time.

It's in this context that our friends in Mission Control are expected to be aligned to our core purpose and perfect in team performance every day. But there are several challenges that make that tough to deliver and a scary role to step in to.

First and most obvious of these challenges: it's still rocket science (and a number of other engineering disciplines). Just to get in the room, there is a wide range of technical material we each must master. This is the book learning – maths, physics, applied engineering etc. The truth is, it's all material that just about any degreed engineer, mathematician or scientist can master. The question is, will they choose to? It is a lot of theory and data to digest and know well enough to apply on the fly in the control room, when you have to get it right because the margin for error is so small. Just at the entry level, it takes a year or more to get up to speed across the technical spectrum, and the learning continues after that for higher-level responsibility on the Mission Control team. Overcoming this challenge alone explains why so many of us chuckle with some pride when we say, 'Why, yes – I *am* a rocket scientist.'

Adding to the challenge, the clock is always ticking, and the spaceship is always hurtling literally to some next decision point. We often run out of time sifting through the data and applying all of the rocket science when the decision is required. If we just had another few minutes – or a day – we could study the problem further. We'd bring in more experts, be 'infinitely smarter', and make a 'bullet-proof decision'. But spaceflight often doesn't give Mission Control that luxury while we're flying. Instead, despite some equipment failures or potential fires in an avionics bay or a myriad of other problems, we are left with dilemmas such as:

- Did the crew really just survive launching into space on columns of fire, only for us to give up and land before we've accomplished the mission?
- If we're going to dock to this space station, we have to burn now, or we miss our window and won't have the ability to try again.
- If we're going to land, we have to do it now, before we run out of gas, air or propellant . . .

Act too soon, without a full understanding of the situation, and we run the risk of over-reacting and making the situation much worse. Or we keep studying, wait too long to take some action, and miss the chance to take the action that would have prevented some bad thing from happening.

What else adds to the scariness of being in this role? The stakes and ramifications of being wrong. In simplest terms, we could blow up the rocket, crash the spacecraft and kill the crew, every day we're on the job. Immersed in rocket science, held accountable to make decisions in real-time, the Mission Control team is always aware that even small mistakes, if uncorrected, can cost us a $500 million mission, a multi-billion dollar spacecraft, or our friends' lives. In fact, this is such a prevalent part of the job that we intentionally remind ourselves of it in The Foundations of Mission Operations, 'To always be aware that suddenly and unexpectedly we may find ourselves in a role where our performance has ultimate consequences.'

Short of those very real stakes, we can embarrass the team, MOD, NASA and the country right on national television. Worse is the fear of letting down the legacy – the generations who came before us in Mission Control and who gave us every opportunity to learn from them to do it right. Although it sounds like small potatoes, not 'measuring up' when it counts is an anxiety that is second only to killing the crew.

To these challenges, we add a last: the human element. This is the host of things that make us personally uncomfortable, all of which just exacerbate the other challenges. We all have them, although they vary from person to person. It's in response to these discomforts that we hear things like:

- 'I already know the answer the boss wants, so why bother?'
- 'That's outside my area of responsibility.'
- 'Oh, that's not how we do that – it can't be right.'
- 'This is the way we've always done it, and it's always worked.'
- 'Who am I to make that call/be the odd guy out/suggest a change/buck the system . . .?'
- 'If I'm wrong, I'm going to get blamed!'

Each of these challenges adds a layer of intimidation and increases the courage required to do the job. Also, like the 'Why We Lead' table from Chapter 1, these challenges are not unique to spaceflight. Every team is confronted with some or all of these, depending on the business and the problems they face.

Because of them, many people shy away from working in Mission Control altogether. Some break under the pressure while training. Those who make it onto the team are faced with overcoming these challenges – the intimidation and fear – to make sense out of the data, stay ahead of the clock and make the right decision, every time, protecting the astronauts and spacecraft, and accomplishing the mission.

In real-time in Mission Control when we're flying, there are no more training runs and simulations to catch errors and weakness in our decision-making and plans. In the most critical emergency situations, there are no more teams of engineers to review some analysis or decision, check our work and offer new ideas. In every moment, each member of the flight control team must be prepared in their area of expertise, overcome the challenges and then step up and make the right decision.

- Several challenges can stack up and increase the courage required to make many decisions, and in some cases can intimidate us into inaction:
  - Technical complexity.
  - Incomplete data before a decision is required.
  - The stakes and ramifications of being wrong.
  - The 'human element'.
- Each member of the team must be prepared in their area of expertise, overcome the challenges, and then step up and make the right decision.

What Do We Do About It?

Talk the Talk

It is a fact of life in Mission Control that our most critical work is difficult and scary. As Gene Kranz reminded us in 1965, that means we must learn to be 'tough and competent'. The trick is to *become* tough and competent, not just say it. It is to overcome the challenges that make the job difficult and scary, because we cannot tolerate being wrong, and we cannot tolerate being intimidated into inaction.

We start with indoctrinating new Mission Control employees, and we're not shy about it. We have formal introductory training with an express purpose to convey our history, responsibility and values, and this training is heavily centred on The Foundations of Mission Operations. New employees are told from the start by supervisors and by the 'old guard' who are already on the job, 'This is who we are . . . MOD, tough and competent. We're the conscience of human spaceflight'.

It's a steady drumbeat that reminds us of what it takes to be in Mission Control, what we are truly responsible for, and how we will be judged. We go out of our way to emphasize the risk of the 'ultimate consequences' if we don't do our jobs, and do them perfectly.

This indoctrination from day one is all about the alignment to Mission Control's core purpose. Yes, the stakes are high, and we are stepping into a role where we can make the difference between life and death for our astronauts. Many will find that scary, but we're becoming part of the Mission Control tradition. We will not let MOD down, so we will not let the astronauts down. We get it – we will protect the astronauts and spacecraft. We will accomplish every mission. This is who we are and the difference we make. Bring on the training, and get me onto the Mission Control team!

Now we're talking! And just like that, Mission Control's core purpose takes on the air of a moral imperative. The drivers for that emotional pull vary for each of us and include: a personal call to protect the astronauts; the challenge to perform at an uncompromising level in such a high-profile role; the desire to join the long line of highly respected flight controllers; and more. The moral or emotional component further reinforces the alignment to the core purpose, and with it comes a powerful and enabling boost in overcoming the challenges, discomfort and fear that come with the job. This is an intangible facet of flight-controller readiness that many Flight Directors recognize as 'fire in the belly'.

Critical as they are, alignment and attitude alone aren't enough. So while we're stoking the 'fire in the belly' and giving them some of the MOD faith, we train them in our rocket science. This is the book-learning phase, although it's not all books. There's a variety of workbooks, e-learning and small classes. This is where we introduce them to the material they may not have learned in college that they'll need to master to work in Mission Control. Some of this is very technical, such as orbital mechanics, flight dynamics and space radiation. It includes details for how every system is designed and operated, and more excruciating detail in their particular area of responsibility. They learn standard spacecraft-operating procedures as well as emergency procedures. As importantly, they learn *why* each procedure is performed the way it is and *why* we consider it the most effective and safe.

As they gain technical expertise, they start learning about the Mission Control Center systems they'll use when flying. This starts as simply as logging into the computer and voice systems, using the video, planning and data play-back systems, and more. These are the tools of the trade that they'll use while applying their technical expertise and flying the spacecraft.

Throughout this process, flight controllers are tested in written and practical exams to ensure they have acquired the depth of knowledge necessary before they can participate on the team. This leads to simplified simulations, working in small teams in limited scenarios, and then to full Mission Control team simulations. After passing a series of progressively more complex evaluations they are sent to the Mission Control Center (MCC) for final evaluation and real-time operations, and that's where they'll experience the final fine tuning.

▸ Deliberately articulating and reinforcing the alignment to the core purpose in moral and emotional terms helps stoke the 'fire in the belly' – the internal drive to overcome the challenges, discomfort and fear that could intimidate us away from a necessary decision.

▸ As critical as they are, alignment and attitude alone aren't enough, and while we're stoking the 'fire in the belly' we also train the team in our rocket science.

▶ As importantly, we want them to learn the 'whys', like: why each procedure is performed the way it is and why we consider it the most effective and safe.

Walk the Talk

The big league at last – the new people are ready for the Mission Control Center. They're aligned to our core purpose. They know the rocket science, spacecraft, procedures and all of our Mission Control systems. They're starting to talk like us.

Now it's time to show us they can use all of that when the clock is ticking in full mission simulations that are indistinguishable from real flight. This is when they learn how to behave like a flight controller under severe pressure in the form of those challenges . . . it's a large volume of complex data requiring quick decision-making, when the ramifications of being wrong can be catastrophic and immediate.

That means they can now screw up, blow up the rocket and kill everybody. Although it is only simulated during training, it is exceptionally realistic, and the pressure is on. Astronauts are flying in a spacecraft simulator, which is sending data to the Mission Control Center just like the real spacecraft. The flight control team is in the control room, headsets on, executing mission plans, responding to the astronauts' calls and data from the spacecraft. Every training run looks, sounds and feels like the real thing, except the problems are usually much more difficult than experienced in a normal day in space.

Fortunately, since these are still training runs, no real explosions and deaths are involved (no rockets are blown up in the making of these flight controllers!). Just the same, thanks to the alignment to our core purpose, every mistake or failure in training feels almost as bad as the real thing. After all, we all know we're not in the control room to 'do the best we can and see how it goes'. We're there to *never* fail, *always* protect the astronauts, the spacecraft and the mission, even in training. Besides, there is the potential for other personal consequences. Fall behind, fail too many evaluations along the way, and trainees can wash out of training and be moved into some other support job, or in some cases will have to seek other employment.

As they prepare for their final training runs, rookies are coached on how to monitor the large stream of data and the cacophony of voice loops, talk on the voice loops (the right level of detail, cadence and vernacular), write logs and technical reports, and more best practices and behaviours. The most important of these best practices were codified in 1988 by Chief Flight Director Brock (Randy) Stone in 'The Stone Tablets of Flight Controller Operations'.[10] These represent the view from the leaders in the back of the room (Flight Directors) who are 'large and in charge' in the Mission Control during flight. After decades of training and evaluating aspiring flight controllers, the Flight Directors have seen all the variations. Presuming you know your rocket science, 'The Stone Tablets' are the specific behaviours Flight Directors want to see and will critique when working with rookies.

Want to pass your final evaluation and make it to the Mission Control Room as a team member? Do these things in all simulations, and then keep doing them when you've made it through. Fall short in any of these behaviours, and expect immediate feedback, often terse and in public. That feedback will usually come from the Flight Director, but it's not unusual for it to come also from peers and co-workers who are observing.

The Stone Tablets are the best practices of good flight controllers. When everyone on the team is doing them, we have the best chance of not missing anything and being perfect in team performance. As such, they are the Flight Directors' '12 commandments'. (For years, I referred to these as the Flight Directors' '10 commandments', only to have a workshop participant point out that there are 12. Fortunately, getting that number right wasn't critical to team performance.) This version of The Stone Tablets has been revised slightly to take out some of the NASA-ese.

The Stone Tablets

I. Show up prepared for your job. Understand how your role fits into the day's activities. Be familiar with any unusual or non-standard decisions or procedures for the activities.

II. Listen closely when the customers and stakeholders talk. If you determine that the information doesn't affect your area of responsibility, then you may go about your business, unless it interrupts other critical work and discussions.

III. Condition yourself to react to customer and stakeholder discussions without having to be prompted by your management.

IV. When customer and stakeholder comments affect your area of responsibility, immediately follow their call with a tailored acknowledgment to your management. For example:
   ▷ 'I understand. I will look into it right away.'
   ▷ 'I understand. This is understood and doesn't require damage control.'
   ▷ 'I understand, and I agree with the customer or stakeholder.'
   ▷ 'I understand. I suggest we take the following action on behalf of the customer or stakeholder . . .'

V. Listen closely to your conclusions that are relayed to your customers and stakeholders. If it has been a while since you discussed the situation with your boss, confirm directly to your boss that you're still in agreement. This lets the boss know the message was communicated as you intended, and you are ready to observe the result. When a customer or stakeholder summarizes the message they received, particularly complicated or important ones, again confirm directly to your boss that you're still in agreement.

VI. Minimize distractions and focus when a key customer or stakeholder is talking. Criticality of the issue will dictate your sense of urgency in responding and resuming other work. A good rule of thumb would be to ensure that the normal point of contact is fully apprised before taking action, if waiting is possible.

VII. An ideal exchange between a key customer or stakeholder and your office is one in which your boss does not need to help. In this case, the boss just needs to be aware of the issue and your appropriate response, and can focus on other priorities.

VIII. Minimize informal or back-channel discussion for key decisions and actions. Use office-wide communications unless impractical to do so.

IX. Be ahead of the customer in finding necessary changes to the project plan based on strategy changes or activities that have not gone as expected. Look ahead.

X. Be as specific as possible when recommending action to be taken to avoid misinterpretation.

XI. If you have time to communicate in writing in some archived way, write it down.

XII. Written or vocal actions should include what you want done, by when, and why.

Based on the Flight Directors' 'Stone Tablets of Flight Controller Operations' by Randy Stone, Chief Flight Director ~1988

A similar but longer summary of best practices was compiled *by* flight controllers *for* flight controllers. Where The Stone Tablets is the Flight Directors' commandments, 'The Autonomous Flight Controller'[11] is the working-level's expectations of each other and the newbies who seek to join them. Like The Stone Tablets, it's intended to spell out how to behave like an expert flight controller who is always in control.

The table of contents is shown below, although the full text is available online (http://www.atlasexec.com/resources/useful-material/), both in its original and de-NASA-esed versions.

The Autonomous Flight Controller

BASED ON 'THE AUTONOMOUS FLIGHT CONTROLLER,'
DON BOURQUE, GUIDANCE, NAVIGATION
AND CONTROL, 1981

CONTENTS

A. Foundations of a Systems Flight Controller

1. Flight Systems Knowledge
2. Ground Systems Knowledge
3. Communications Ability
4. Initiative/Attitude
5. Aptitude

B. How to Lead Your Leader

1. Understand that your leader is not very smart
2. Your leader is lonely

C. Flight Systems Management Techniques

1. Systems Active Reports
2. Problem/Failure Reports
3. Retracing Events
4. Identify Cause, Scope and Effects
5. New Workarounds
6. Get Next Failure Analyzed

D. Basic Disciplines

1. Communications
2. Documentation
3. Area
4. Team and Self

A final example of this kind of reminder is found in the 'Links of the Error Chain'. The precursor to this list was originally compiled as a result of aviation mishaps over the years. They are the typical factors that contribute to pilots missing some cue and making an otherwise avoidable mistake. Any one of them adds risk. When things aren't going well, the situation can quickly erode into multiple of these links being present, which is a recipe for disaster, in many cases avoidable disaster.

The Links of the Error Chain[12] shown below is a de-NASA-esed version, while the original cue card many flight controllers keep on console is available online (http://www.atlasexec.com/resources/useful-material/).

Links of the Error Chain

- ▶ Failure to Stay on Plan.
- ▶ Executing Undocumented Procedures.
- ▶ Departure from Standard Procedures.
- ▶ Violating Limitations.
- ▶ No One Looking at Data and Metrics.
- ▶ No One Listening to Customers and Stakeholders.
- ▶ Poor Communications.
- ▶ Ambiguity.
- ▶ Unresolved Discrepancies.
- ▶ Preoccupation or Distraction.
- ▶ Confusion or Empty Feeling.

Break the chain, now!

As a Flight Director, I kept this cue card on the console in front of me and regularly glanced down the list to assure myself none of these were creeping in.

In many training runs, after we have experienced multiple system problems, it is not uncommon to have two or three maintenance and recovery procedures in work by different groups of astronauts in different areas of the spacecraft. As each group calls down with different questions or takes some action in fixing our problem, it is easy to mix up the calls and actions for the separate procedures. For some of the more complicated and lengthy procedures, you could find yourself wondering, 'I wonder where the hell we are? Are we sure the electrical system recovery is ready for the next steps in the computer system recovery? Does my communications expert know for sure what we're doing, and is he confident we won't lose connection with the spacecraft in the next step?'

All great questions. Any one of them is a warning sign that

we're at risk of making a mistake or missing something important. Also, any one of them tells us it's time to focus and eliminate a link in the error chain before it's too late.

If I had any of those thoughts, or got the impression that anyone on my team had them, I'd stop work in the control room and onboard the spacecraft, talk to the flight controllers and astronauts as a group, and quickly get us all on the same page: our progress in all procedures so far; the intended outcome of each procedure; overlap or impacts to other systems; and our communications in our remaining actions etc. Then we'd start back to work, taking deliberate action on the spacecraft with the full team completely aware of where we are and where we're going – me included.

Of course, bullet lists like The Stone Tablets, The Autonomous Flight Controller and the Links of the Error Chain are great crib sheets while training to turn the behaviours into habits. They also help preserve a sense of calm when we're under pressure responding to real-time problems and emergencies.

Regardless of how well trained, we will all still feel the pressure from the challenges that make the job difficult – to make sense out of complexity, stay ahead of the clock and make the right decision. As the pressures build in real-time, the adrenaline starts to flow, and we're at risk of forgetting our training. This can lead to tunnel vision and then to panic, neither of which leads to good decision-making. A quick mental inventory of these top-level best practices and behaviours can reassure us we're working towards doing the right thing. Or, if we realize we're neglecting some of those best practices and are now seeing links in the error chain, we are reminded to make a deliberate extra effort to do what we've learned. Otherwise, the risk of failing is increasing all around us, we may not be doing anything to stop it, and we may just be making it worse!

Through this indoctrination and training process, we progress from embracing the core purpose and 'talking the talk' to *walking* the talk – habitually *behaving* like flight controllers. All of the tips, best practices and expected behaviours become second nature through practice in training and experience on the job. In all decision-making, we demonstrate the behaviours from The Stone Tablets and The Autonomous Flight Controller. We remain sensitive to, if not paranoid about, The Links of the Error Chain and the continuous need to eliminate them. We have been prepared to the full extent of Mission Control's experience to control our destiny, not to be victim of it, or – more accurately – that our astronauts will not be victim of it.

We're ready now to be in the Mission Control Room, protect the astronauts and accomplish the mission.

▸ Training is critical not just for the book learning but also to progress from embracing the core purpose and 'talking the talk' to walking the talk – behaving like flight controllers. This includes:
  ▷ Embracing the responsibility of always achieving our core purpose.
  ▷ Applying tips, best practices and lessons learned from previous experiences.
  ▷ Demonstrating both the knowledge and behaviours consistent with the best practices while under pressure.
▸ We are ready for the responsibility of our core purpose when we know our rocket science and demonstrate these best practices and expected behaviours as second nature:
  ▷ In all decision-making we demonstrate the behaviours from The Stone Tablets and The Autonomous Flight Controller.
  ▷ We remain sensitive to, if not paranoid about, The Links of the Error Chain and the continuous need to eliminate them.
  ▷ We have the 'fire in the belly' to engage on anything we see and hear, apply what we've learned, and make a deliberate call – to make a difference.
▸ Through this process, we are prepared to the full extent of our experience to control our destiny, not to be victim of it, or – more accurately – that the people relying on us will not be victim of it.

CHAPTER 3

Trained to Be Trusted

*Everything should be made as simple as possible,
but not simpler.*

Albert Einstein[13]

Why?

One last step remains to progress from indoctrination and trained behaviours to understanding the deliberate real-time morality. A hint about that final step can be found in the continued growth flight controllers exhibit during final training and on the job. Beyond the 'indoctrination' and formal training, as we gain experience working in this environment, flight controllers and Flight Directors become natural inquisitors, quick thinkers on our feet, and concise and unhesitating communicators. If something happens on the spacecraft, even something small, we are immediately asking and assessing:

- What just happened and why?
- Did we expect it? If so, did it happen as expected? If not, how was it different and why?
- Do we have indications of a failure, either some piece of equipment, its power supply, controller or instrumentation?
- What evidence do we have to support any of those answers,

and can we trust it? Is there an unrelated failure that could be misleading us in any or all of them?
- Is there any risk to the crew, the spacecraft or our current operation? If so, what are our options to respond, and what is the risk of each?
- What action do we recommend and why?

Why, why, why? And why isn't the technical training enough? Why not just teach them the rocket science, system designs and procedures and make sure they always follow them? Because that kind of training and the reflexive (Pavlovian) button-pushers it produces are only as good as the procedures and the data. Any errors in the design information, software or manufacturing can result in a system that does not operate the way it was intended, or as it was written in some procedure. As systems become more complicated, they can fail in unanticipated and more complicated ways than envisioned when operating procedures are first developed. In the end, if the expected warning light doesn't turn on, our 'trained' operator may not take any action as systems fail. After following a procedure as written (and as trained) and finding the spacecraft still not functioning, this kind of 'trained' operator has nothing left to offer.

....

That's Not Supposed to Happen. Now What?

As an example, take our unanticipated experience on 17 March 2001 while flying Space Shuttle Discovery. Nine days into STS-102, the Discovery and International Space Station astronauts had finished two spacewalks and were nearly finished resupplying ISS. The astronauts and Mission Control were already preparing for Discovery's undocking in two days and landing two days after that.

Around 3 a.m. in Houston, a Shuttle flight controller (EECOM) called me (Flight) on our primary voice loop to report a problem with the cooling system. After 10 minutes of normal recovery procedures, the situation became critical:

Flight: Where are we with Freon Loop 1 now, EECOM?
EECOM: Flight, we can restore flow to the loop, but not through the radiator. We've already taken all of our normal recovery steps. They did not restore normal operation. That means the Freon that is stuck in the radiator will continue to cool down below our allowable limits.

My concern now is that we have ice in the radiator; it's the ice that's causing the blockage, and it's getting worse – it is on its way to becoming a big slushy. The engineering support team tells me it isn't possible to have ice in the Freon loop. They say it must be something else.

However, if we do have ice, it could flow down the line to the Freon/water interchanger. That interchanger is so efficient that ice in the Freon side would cause the water lines to freeze and crack the metal housing. If that happens, the crack will create leaks that empty *all* water and Freon loops at once. *This could result in the loss of the entire Shuttle.*

[Note: If we had lost all Shuttle cooling while docked to ISS, we would have had to abandon Shuttle and keep the astronauts on ISS for rescue on another mission.]

EECOM: I don't have any other ideas on what could be causing this behaviour. In spite of what Engineering tells me, based on the leak risk, I recommend we assume this blockage is ice.
Flight: Got it. Concur, EECOM.

All flight controllers, coordinate with EECOM on equipment you can power up that will dump the most heat possible into the cooling system to prevent that interchanger from freezing.

EECOM, how about we manoeuvre and point the cold radiator towards a warmer part of the sky to help thaw it out?
EECOM: We were going to recommend it, Flight.

Flight: Rog. FAO, get us an attitude that puts direct sun on this radiator while we have the crew turn on more equipment to heat things up from the inside.

Station Flight, we need your help to coordinate handing attitude control from ISS to Shuttle while we manoeuvre and try to keep this from freezing and killing all cooling.

Station Flight: In work, Shuttle Flight. Let us know what else you need.

Within minutes, the ISS mission control room had coordinated the procedure with the Russian control center in Moscow and with the ISS astronauts. The Shuttle astronauts then activated the Shuttle's attitude control system and started pivoting the million pounds of Shuttle and ISS to point Discovery's frozen radiator at the sun.

Meanwhile, the entire Shuttle flight control team identified equipment to the EECOM to help heat up the cooling loops. The Capcom relayed all of the equipment to the astronauts, who quickly powered on each device.

EECOM: Flight, because of the lower temperatures, in order to keep the Freon flowing, we will have to disable under-temp (freeze) protection. Otherwise, the system will keep turning itself off, and we will not be able to flow even small amounts of Freon through the radiator.

I know it doesn't sound right, since under-temp protection is designed to keep the interchanger from freezing. But if we don't get this Freon loop flowing through the radiator, we will lose it and have to deorbit right away. Running the Freon loop without thawing the radiator risks freezing the interchanger anyway and then losing all cooling, which is likely not survivable.

Flight: Copy EECOM. I concur. You're go.

EECOM: Flight, let's have the crew power on both water loops to keep the interchanger and both water loops as warm as possible. Then recycle the Rad Controller one more time, and switch the Bypass Valve to manual (thereby removing under-temp protection).

Capcom relayed the instructions, and the crew jumped on it.

Twenty minutes later, normal flow resumed in Freon loop 1, and it was quickly back to normal operating temperatures. An hour after the first indications of a problem, Shuttle had manoeuvred the space station back to its normal attitude, and the astronauts had powered off all of the extra equipment we relied on to protect the interchanger from freezing. It was like nothing had happened.

After the mission, the Kennedy Space Center (KSC) discovered three times the allowable level of moisture in both Freon loops. They also performed tests on Freon pumps based on the confusing electrical signatures of the pump as the line blockage formed. The tests confirmed that the community's decades-long expectations and our flight controller training were exactly wrong. The electrical signatures with stopped-up lines looked identical to what EECOM had seen in flight. KSC had confirmed that ice-blockage in the radiator was the most likely culprit.

That hour of spaceflight was a thing of beauty. EECOMs Mike Fitzpatrick and Curtis Stephenson (Mike's trainee) were on top of this system failure from the start, from their response to the initial Freon loop temperature drops and radiator flow problems to Mike's out-of-the-box thinking and dogged pursuit of ice in the system. With the recognition of the enormous risk the Shuttle and its crew faced, the entire team responded like we had rehearsed this a hundred times.

When it was over, ISS Program Manager Bill Gerstenmaier walked into the Shuttle control room grinning from ear to ear. He sat down next to me at the Flight Director console, 'Man, I listened to that whole thing in the conference room downstairs. It was incredible. Incredible. You guys are so good.'

....

And *that's* why we focus on more than just training the 'correct' response to some stimuli.

We're not there to plough ahead when something doesn't make sense, stubbornly (or blindly) sticking to a plan and procedure that could be wrong. There's no partial credit for simply taking the action we were taught until we give up the mission or crash. Each of us is groomed, first deliberately in training and then through experience on the job, to ensure we will unhesitatingly make the right decision – do the right thing – every time.

We judge ourselves and each other on both being right and in getting there the right way. The managers and stakeholders outside the Mission Control Room pass judgement on the team in the same way: 'Are they making the right decisions? Can we trust them to protect our astronauts, protect our spacecraft and accomplish our mission?'

Trust. Now we're getting to it.

'In God we trust, all others bring data.' That sign hangs over the entrance to the Mission Evaluation Room (MER), a room in the MCC that houses engineers who reinforce the analytical horsepower of the flight control team while we're flying. (Among other important roles, the MER is the first place Mission Control calls when we need help finding and analysing data.)

If the Mission Control Room had an equivalent sign, it would read, 'In God we trust, all others must keep earning it.'

From the day we start, we are really being trained to be trusted. Everything from the alignment to core purpose to learning to walk and talk like flight controllers is geared towards developing individuals and teams that can be trusted implicitly to:

- Sift through the complexity.
- Quickly evaluate the situation and options.
- Remember our training and the 'corporate experience bank'.
- Overcome the intimidation and fear.
- Then be perfect in decision-making and action, which

includes realizing when the procedures and our previous experience are misleading us.

Stephen M.R. Covey sums up trust succinctly in his book *The Speed of Trust*[14]:

> Trust is a function of two things: character and competence. Character includes your integrity, your motive, your intent with people. Competence includes your capabilities, your skills, your results, your track record. Both are vital.

While still important, *competence* is the easy one. This is the book-learning component of our training. Learn the rocket science or don't. This is easy to measure: pass the evaluations throughout training or don't. But when you've made it through and are certified to sit in Mission Control, you will know your stuff. You will be *competent.*

The key to leveraging the 'fire in the belly', applying the book learning and doing the right thing in Mission Control under pressure is the specifically cultivated *character* component of trust. Beyond competence, this character component is the underlying theme in The Foundations of Mission Operations, The Stone Tablets, The Autonomous Flight Controller and the Links of the Error Chain. Throughout the indoctrination and training in these behaviours, it is this underlying character component of trust that we want to see and reinforce. The fundamental elements of the character component of trust in Mission Control can be distilled down to: technical truth, integrity and courage.

▶ From the alignment to core purpose to learning to walk and talk like flight controllers, our goal is to develop individuals and teams that can be trusted implicitly, especially under pressure, to:
  ▷ Sift through the complexity.
  ▷ Quickly evaluate the situation and options.

▷ Remember our training and the 'corporate experience bank'.

▷ Overcome the intimidation and fear.

▷ Then be perfect in decision-making and action, which includes realizing when the procedures and our previous experience are misleading us.

▸ The key to leveraging the 'fire in the belly', applying the book learning and doing the right thing under pressure is the specifically cultivated character component of trust, comprised of: technical truth, integrity and courage.

Technical Truth

This element of trust appears obvious and easy, and it should be. In Mission Control, we are expected to base our decision-making on reality. 'Technical truth' means what is *really* happening, not what we had planned or what we hope or wish was happening, but the real, no-kidding situation.

A number of things make that difficult. For starters, the universe has a way of not behaving according to plan. In spite of our extremely well thought-out and rehearsed plans, 'stuff happens'.

Some things take longer to do than we predicted when laying out our plan. As they do, the choreography between things like satellite communications coverage, robot arm movement, space-walkers etc gets all out of whack. If we're not paying attention, we can quickly end up in an unplanned predicament, such as having astronauts waiting around in their spacesuits to finish a spacewalk because we didn't move the robot arm into place on time. While they wait, they're cooped up in the spacesuit, using up the limited ability the suit has to clean carbon dioxide out of the air they're breathing, and the ISS may fly out of range of the satellites that had been scheduled to provide communications during the critical part of the spacewalk. Not good.

Even when things appear to be going as planned, we can overlook some detail when making a plan that stops us in our tracks. For example, we may plan to activate some equipment but overlook a secondary computer that is required to control it. Or, to further complicate matters, we may have different and incompatible versions of software in two computers that present a problem only when we attempt today's operation. Correcting it could be as simple as reloading the computer with a different version of software, but depending on the computer, that could take days to test and transmit to the spacecraft.

If we're able to catch any overlooked details and stay on time, equipment can act up or break in more ways than we can count – and not always in the way we expect. This doesn't have to be big failures, like engines exploding. Seemingly small failures can bedevil the best laid plans.

For example, some smoke detectors can set off false alarms when astronauts are nearby moving things around. The movement stirs up dust which looks like a cloud of smoke to the sensitive smoke detector, and a fully normal smoke detector can set off a false alarm. In other cases, a smoke detector can break in a way that causes it to spontaneously signal an alarm when no smoke is present. That's why we look for multiple indications before taking some extreme action, such as shutting off ventilation and dousing the fire with Halon, neither of which is good for the astronauts' breathing. That's also why we didn't immediately abort the rendez-vous described in Chapter 1 with only a single smoke detector going off.

And just like that, we experience the first of the Links of the Error Chain from Chapter 2, 'Failure to stay on plan'. Realizing we're falling behind, the pressure starts to rise from the challenges that make things more difficult. We become more aware of the weight of the technical complexity, wanting more time to think and get help, the ramifications, and the human element. And that pressure can push us to answer quickly, as if the quick answer makes the problem go away. As we give in to that impatience, we may continue working our way into more and more Links of the

Error Chain. We then find ourselves at increasing risk of making uninformed decisions and guessing.

We flower up our guesswork by calling it things like 'engineering judgement', 'intuition', 'educated' guessing and 'trusting my gut'. In other settings when we don't feel the need to justify our lack of rigour, we'd call it 'shooting from the hip' or 'spit-balling'. In all cases, it's guessing. It's one thing to guess our way through putting all of the options on the table for discussion. However, in real-time, when we are making a decision and will take action, whether we find ourselves adding Links of the Error Chain because we fell behind or we lost discipline in our thinking from overconfidence, boredom, laziness . . . the end result is the same.

If our guessing is correct, we may get away with our decisions. But that 'success' will not be because we have technical truth. The longer we go like this, the further we erode our grasp of technical truth and are simply relying on luck.

It is the *reality* that can quickly get out of control and kill the crew when our guessing and poorly thought-out decisions lead us astray. Much of our training and the habits we learn from The Stone Tablets and The Autonomous Flight Controller reinforce our awareness of the Links of the Error Chain as they creep in. It is our focus on technical truth that eliminates those links, increases our grasp on reality, and helps us succeed, not through blind-ass luck, but through deliberate action.

▶ Technical Truth:
  ▷ Seek and base decision-making on reality – the actual – not expectation, hope or desire.
  ▷ This focus helps eliminate the Links of the Error Chain, increases our grasp on reality and helps us succeed by deliberate action.
▶ It is the reality that can quickly get out of control and end in failure when our guessing and poorly thought-out decisions lead us astray.

Integrity

If only our decision-making could always be as easy as simply waiting until we have all of the data, analyses and discussion necessary to have full situational awareness – full technical truth. Then we could make the elusive bullet-proof or no-brainer decision every time. There's no risk of error, so there's no pressure.

As we've already seen, though, the universe just doesn't work that way. It is frequently difficult to have enough data to fully understand a situation, to have complete technical truth. But the clock keeps ticking, and we just as frequently run out of time and are forced to make a decision, to pick a fork in the road, because like it or not, the spacecraft is still hurtling through the sky.

What we demand from each other and from the team in every decision is an unrelenting *pursuit* of technical truth and an awareness of the parts of our deliberation that are less supported by data and real knowledge. We are expected to retain a deliberate, ongoing awareness of what we *know* versus what we *think*.

Thus our obsession with our series of 'Whys?' at the beginning of this chapter. It isn't enough to have an opinion, to make a recommendation or to take an action. Surviving our training, being declared an expert, and having a position or title doesn't grant us omniscience. Why are we right? On what is each opinion based? Do we have data to back it up?

If we don't have all of the data to prove our point, can we describe the engineering principle or theory underlying our recommendation? Are we able to compare the risks of different options and weigh them against the uncertainties in our knowledge? In other words, if we're wrong, what is the worst thing that can happen? Is that most likely for the part of our knowledge that is weakest, the part that is supported by the least data, the least technical truth?

The integrity we seek from every flight controller and from the team isn't an esoteric reminder to be honest. Integrity in Mission

Control is our conscience in pursuing technical truth. It is that continuous drumbeat and paranoia that we're missing something. Like our focus on technical truth, this integrity is reinforced through our training, and the habits we learn from The Stone Tablets and The Autonomous Flight Controller reinforce our awareness of the Links of the Error Chain.

When evaluating data and taking a position, flight controllers learn to ask and re-ask some basic questions:

- Have we accepted an easy or superficial explanation that hides what is actually happening?
- Why do we think we know something?
- Based on what measurement, analysis, experience or judgement (guess)?
- How much uncertainty is there to our knowledge of technical truth, and why?
- How does that uncertainty change what the actual situation could be onboard the spacecraft and in the environment?
- What are the worst ramifications of the most probable of the uncertain alternative situations, and how do those ramifications affect our decision-making, risk assessment and *action*?
- Have we learned anything new that could change our assessments since beginning to work on this problem?
- Repeat, time permitting . . .

Like so much of Mission Control's culture, this integrity for technical truth has been present and growing since the beginning.

You can see it as Chris Kraft describes mission rules in his 2005 lecture at MIT[15]:

> Now, the other thing that we invented at that time – I say 'invented'; it just came about by evolution – was a book called Mission Rules. And that was probably the smartest thing we ever did.

As we began to look at the spacecraft systems, we started asking questions.

- If this system is failing, what are the measurements that we're going to have there?
- And, if it is failing and it isn't operating at the right temperature or the right pressure and it is off nominal, what will the system do?
- And how do we measure that on the ground? How do we detect it?
- Where is the instrument located on the system? Because it might be affected by the position it is located in the spacecraft.
- It might be hot. It might be cold.
- It might be suffering different kinds of pressures than [when] it was measured on the ground.

And, as we began to ask those questions of the system engineer – the 'system' engineer, not 'systems' engineers, because I don't think we had any at the time – and they would say, 'Why the hell do you want to know that? The system is either working or it ain't working.'

And we said, 'Yes, that's a good answer, except that now we've got this system in space. And if we want to continue this flight and not have a contingency operation, we would like to know how long the system is going to last if it isn't operating under normal conditions.'

That prompted us then to start thinking about how the system failed and what we were going to do about it.

- If the thermal system that kept the astronaut from getting hot or getting too cold wasn't functioning properly, what could we do about it?

- How long could he stand being at a temperature of 85 degrees inside his space suit?
- And then, that said, 'Well, if it stays there, and we can only go X number of minutes, what are we going to do about it?'
- What is the rule of the game that says we should re-enter or not re-enter or go to the next primary recovery area, et cetera?
- And it allowed us then to write down, for every system, and understand what we would do under certain circumstances. [We] called those a set of 'Mission Rules.'

. . . It began to have everybody start thinking about how does my system fit with everybody else's system? How does that fit with the game plan that we're trying to come up with?

Mission Control turned this kind of thinking into a science unto itself. Over the years, we evolved past just the flight rules (as mission rules came to be known in the 1980s) and put just as much effort into documenting the 'whys'. Many of the situations described in flight rules were complex, and made more so depending on equipment failures. This complexity often required weeks, and sometimes months, of engineering study, simulations and hardware tests to fully understand all facets of a problem. Those were then followed by a series of technical reviews to digest the new information and map the best path through a series of hard choices that gave us the best chance to protect the astronauts, the spacecraft and the mission.

It is not uncommon to have a straightforward, one-or-two-line flight rule followed by several paragraphs and sometimes a page or more of detailed explanation for *why* we chose this logic as the optimum, which we called the rationale. The excerpt from a Shuttle flight rule[16] below is an example.

A2-207   <u>LANDING SITE SELECTION   (CONTINUED)</u>

D.   SYSTEMS FAILURE PRIORITIES [1] [3] [5] [6] 6[102402-5804C]

FOR SYSTEMS FAILURES THE LANDING SITE PRIORITY SHALL FOLLOW
THE EOM PRIORITIES. CONSIDERATION WILL BE GIVEN TO THE
FOLLOWING EXCEPTIONS:

SYSTEM FAILURES [1]	PRIORITY	
1 OR 2 APU/HYD SYSTEMS	LANDING SITES WITHIN SINGLE APU WX PLACARDS	[2]
2 IMU'S	EDW 22/04 NORTHRUP KSC	
2 NORM/LAT AA'S		
2 RGA'S		
2 ADTA'S	LANDING SITE WITH MOST FAVORABLE ATMOSPHERIC CONDITIONS	[4]
2 FCS CHANNELS (SAME SURFACE)	NORTHRUP EDW 22/04 KSC	

Excerpt from a Space Shuttle flight rule [NASA]

This short, tabular flight rule is intended to reflect how Shuttle runway-selection priorities change depending on previous equipment failures that affect the Shuttle's flyability and control precision. Four pages of detailed rationale, including runway lengths, surface materials, lighting, ground firmness around the runways, proximity of civilian populations, and more, follow this succinct table. This was one of hundreds of flight rules covering every system and the full operating environment. The Shuttle flight rules alone comprised a book that was 4.5 inches thick, most of it rationale. A separate, similar-sized volume holds the ISS flight rules.

In all cases, each flight controller is expected to know and understand the rationale for every flight rule that applies to their area of expertise and responsibility. In general, Mission Control is expected to fly the spacecraft within the bounds of the flight rules. Further, each of us is expected to understand and correctly apply the intent of the flight rule, which we learn through the rationale and the more detailed records of the engineering discussions that settled the flight rule in the first place.

However, we are also expected to realize when we are in a situation that isn't specifically addressed in the flight rules, or some unforeseen 'exception to the rule' and then do the right thing. As mentioned several times already, we are then expected to protect the astronauts, spacecraft and the mission, even if that means flying outside the guidance of the flight rules. As in all other aspects of our decision-making, we then expect the team to understand *why* the flight rule does not apply in a particular case, *why* a recommended action does not increase risk to the astronauts and is the *right thing to do.*

Again, why, why, why? And all of it is intended to focus the decision-making on technical truth, retaining a deliberate focus on what we know versus what we think, and understanding why we reach every conclusion.

This method of thinking, forming decisions and recommendations, and taking action is groomed into us through training. This is integrity as a trust element in Mission Control. It is what keeps us honest in our thought process. It is our conscience in seeking technical truth. As described in Chapter 2, this integrity drives us to remain sensitive to, if not paranoid about, The Links of the Error Chain and the continuous need to eliminate them.

▶ Integrity:
  ▷ Continuously challenge the boundary between technical truth and assumption – what we know versus what we think.
  ▷ This is our conscience in pursuing technical truth, ensuring we know why we reach each conclusion.

Courage

No guts, no glory, right? Not exactly.

Mission Control isn't after glory, and we don't want false bravado polluting our thinking in critical situations. Instead, we demand individual and team decision-making that reflects this specific integrity in pursuit of technical truth, all aligned to our core purpose. As we've seen, in real-time, while the spacecraft and our astronauts fly through the sky, those demands come with pressures and often downright fear. Channelling my inner Yoda, I have told more than one gung-ho new Flight Director who professed not being afraid of their new responsibility, 'You should be.'

That isn't for dramatic effect. This is serious business, with real and potentially catastrophic consequences for every action we take. We can blow up the rocket, crash the spacecraft and kill everyone. It *is* scary. We don't do ourselves or the team any favours by pretending otherwise.

To compound the pressure as we work through every normal or crisis situation, we're wearing headsets for our peers, managers, stakeholders and sometimes the world to hear our every utterance. We are surrounded by veterans and, in many cases, heroes of the business. They're all experts in their own right who judge us on both our technical ability and demonstrations of those habits we learned in training and summarized in The Stone Tablets, The Autonomous Flight Controller and The Links of the Error Chain. Most days in the Mission Control Room feel like an oral exam in graduate school.

Not only are we always at risk of failing but also of not measuring up, even if we don't fail!

As the pressure builds, we are at risk of being intimidated out of being tough and competent. But we get no credit for our strong Mission Control integrity, pursuing technical truth or being the smartest rocket scientist in the room unless we muster the courage to speak up. In our management ranks, we refer to this courage as 'the willingness to take personal risk'. This isn't the same personal risk the astronauts take when they strap themselves into the spacecraft and trust us with their lives, but it feels just as real after our 'indoctrination' and alignment to core purpose.

This courage is the 'fire in the belly' we are intentionally stoking, the internal drive to overcome the challenges, discomfort and fear that could intimidate us away from understanding and action. While we acknowledge the pressures and intimidation, it is the courage to overcome them in this setting that is the final crucial element of trust in Mission Control. It is this courage that ensures we will engage on anything we see and hear, apply what we've learned, and make the call – make a difference.

▶ Courage:
  ▷ Acknowledge the pressures, intimidation and the risks, and then take deliberate action aligned to this integrity.
  ▷ It often takes courage just to do the job, and it takes still more to adhere to the values – to the integrity – especially under pressure.

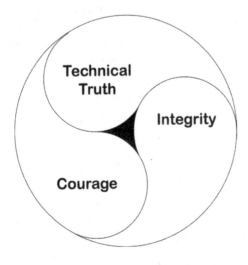

Mission Control trust elements and performance expectations

In the most critical real-time moments of human spaceflight, the room for error and the time available to take action are both very small. The crucial element of Mission Control's unwavering effectiveness in those moments is this very specific trust. To maintain that trust, everyone on the team shows up having done their homework and already being technical experts. They then must demonstrate an unyielding search for technical truth and have the courage to make the call.

Each of the trust elements reinforces the next. Mustering the courage to take on the responsibility also helps us speak up in spite of the intimidation in adhering to this judgemental integrity. The integrity compels us to challenge what we know versus what we think, to justify why we reach every conclusion. That in turn helps us anchor our decision-making in technical truth. As we work our way through each next problem this way, each experience strengthens our resolve, boosts our courage and aligns us more rigidly in our integrity and focus on technical truth.

It is this hard-earned trust, as defined by technical truth, integrity and courage, that is the key enabler in Mission Control.

▸ Mission Control trust anchors decision-making in technical truth, retaining a deliberate focus on what we know versus what we think, understanding why we reach every conclusion, and then stepping up and making the call.
▸ Each of the trust elements reinforces the next.

It Is a Morality

Together, these comprise the real-time morality: unyielding alignment to purpose and deliberately applying the Mission Control trust elements of technical truth, integrity and courage in all decision-making.

Make no mistake; in the control room it *is* a morality. Real people trust us with their lives in these spacecraft. That comes with a very real sense that we would also be failing their families and the larger NASA community who are outside the Mission Control Room, all of whom are also trusting us.

While intimidating, there is power in the emotional charge that comes with that awareness. It strengthens our alignment to our core purpose and helps put the steel in the 'steely-eyed missile men and women' who have to make the calls. This is that internal spark we get from a series of realizations:

- This thing we do matters.
- What I do as part of it matters.
- What I do can have the ultimate consequences for someone.
- I can make a difference.
- I must make a difference.

Chris Kraft referred to this in his book *Flight*, when he described the mood after the Apollo 1 fire that took the lives of astronauts Gus Grissom, Ed White and Roger Chaffee. 'This determination to make sure these men did not die without cause, I believe, gave us all the strength to continue our job of landing men on the moon. It also brought us all closer together and made our responsibilities crystal clear.'[17]

Even with that clarity, however, the need to do *something* doesn't justify doing just *anything*. Our goal is *deliberate* decision-making and actions aligned to our core purpose, instead of panicking, giving up or guessing. Without a full understanding of a situation and our actions, we may get lucky and make it through, but we wouldn't consider the performance a success. We'd still be guilty of not *deliberately* managing the risk. We are guilty of not basing our decision-making on technical truth; thus, we are guilty of breaching the real-time morality.

► Together, these comprise the real-time morality: unyielding alignment to purpose and deliberately applying the Mission Control trust elements of technical truth, integrity and courage in all decision-making.
► It is a morality – there are real consequences for every action we take, from financially costly to catastrophic and life threatening.
  ▷ The emotional charge that comes with that awareness strengthens the alignment to our core purpose and helps put the steel in the 'steely-eyed missile men and women' who have to make the calls.
  ▷ The emotional charge can also make the difference in tough situations to retain the discipline to take deliberate, well thought-out action, instead of panicking and not managing the risk, not basing our decision-making on technical truth.

Subtle Errors

The real-time morality doesn't help us only in crisis, when we're at obvious risk of blowing up the rocket. Consider the hundreds of 'less critical' decisions individual flight controllers make each day as they perform the wide range of tasks required in a normal day of human spaceflight (from Chapter 1). From incremental changes to the day's operating plans to the thousand computer commands uplinked daily to the ISS, even without equipment failures and emergencies, there's an endless cascade of decisions and actions that must still be done well. Small errors in any of these that aren't corrected by the team could be the 'fly in the ointment' that leads to some next crisis.

....

'Better' Sometimes Really Is the Enemy of 'Good Enough'

Mission Control was reminded of this on 24 April 2001 when resetting the International Space Station's command and control computers in response to a problem with one of the hard-drives. This was a procedure that had become almost routine based on how often these hard-drives required resetting. However, because the team was falling behind on the day's schedule, they simply switched to one of the two back-up computers and skipped some time-consuming steps to prepare the remaining back-up computer. These steps were planned for later in the day, after the team had caught up.

As they resumed other work, they saw signs of the same hard-drive problem on the computer they had placed in control. Although this computer was in adequate condition to support the day's activities, the Flight Director chose to switch to the remaining back-up computer. Since this last computer had not been prepared, it did not activate, and instead Mission Control lost contact and control of the ISS.

Fortunately, Space Shuttle Endeavour was docked to ISS, giving Mission Control a temporary back-up route to send commands to ISS to fix the problem. The astronauts rerouted the network cables inside the ISS, patching some older computers directly to the primary control network. This gave Mission Control the ability to slowly retake control, restore all three C&C computers, and then resume normal operations.

The recovery was masterful work, and this crisis was averted.

This was a potentially vehicle- and mission-ending crisis that resulted not from some catastrophic failure like an explosion. It came as a result of a small oversight in a lengthy and routine procedure, while responding to a non-threatening periodic computer glitch and being too aggressive to try to make an already stable situation better.

....

Again, small errors that aren't caught and corrected can lead to severe crisis. In every decision, even the seemingly benign, each flight controller must still overcome the challenges from Chapter 2: technical complexity, incomplete data before a decision is required, stakes and ramifications of being wrong, and the 'human element'.

In doing so, it is rarely a maths or rocket science error that trips up Mission Control. It is a range of the 'human element' pressures that typically lead to some error or omission getting by undetected and some decision that falls just short of the real-time morality.

Previous success can give us confidence that sometimes gets in our way. As the investment ads remind us though, 'Previous success is no guarantee of future performance'.

It becomes easier and easier to forget that good advice after we've been successful, particularly in difficult endeavours such as human spaceflight. After all, *we* are the conscience of human spaceflight. When it comes to the routine and simple tasks, we've been there and done that many times, in much scarier situations than this. We know them when we see them and can frequently – in almost all cases – predict the correct actions with only the first hints of data.

Accordingly, if we're not deliberately disciplined, we can give in to 'shooting from the hip', not because we're feeling the heat of a crisis, but because we know how good we are, what we've already accomplished, and that this one is something we clearly recognize and can handle on the fly. Every time we reach an early conclusion that is shown to be correct, we run the risk of having that behaviour reinforced. 'See? We were right. Our experience and success have led us to just *know*, to have "engineering judgement" and intuition – intuition that we may unintentionally rely on at the expense of technical truth.'

This reliance becomes progressively easier with each successful outcome. And each time we are at risk of adding Links in the Error Chain.

This becomes more tangled as we involve respected peers and leaders, many of whom are regarded as heroes of our business for some specific accomplishment – or a whole string of them. These are experts and leaders who are known to be good and who are inherently trusted with solving our most critical problems. They can carry the E.F. Hutton power – when they speak, people listen. Where we can get in trouble is, when the heroes speak, others may listen but not second-guess them: 'Well, if that guy says this is the answer, it must be right. He thinks faster than anybody, and he's always right. Don't you know who he is?'

Guess what else? Those heroes know that's their status, too. Whether intentionally or not, they can sometimes also fall victim to the justifiable confidence in their own judgement, to the extreme of not fully thinking through their decisions. Like the rest of the team, they may relax in second-guessing their own thinking, in checking the integrity in their decision-making, because they 'just know'. They may start to buy in too much to their own myths.

As a young Flight Director, I had spent weeks trying to solve a complicated challenge with installing a new piece of ISS equipment on the outside of the spacecraft. This was something that I had years of experience in, and detailed systems knowledge. A senior manager, an exceptionally accomplished space engineer with considerable Shuttle expertise – but no depth in ISS systems – suggested: 'Just tell me about the problem. *Even though I don't know the details*, you know I'm good at this stuff. I've learned that I have an ability to solve problems subconsciously. *I don't even have to think* about them consciously. A few days from now, the answer will just come to me.'

That's an exact quote, and he was dead serious. He hadn't always thought that way. By the time he said it to me, though, he had definitely become one of the 'heroes', and he thought so, too.

Again, the risk goes up that we're adding Links in the Error Chain.

Finally, cultural norms can magnify this problem. Whether it's because 'we've always done it this way', or the boss has decided, some hero has already weighed in, the team has consensus . . . something has set a precedent, and the issue has been declared solved. Now that it's gone through the process, we trust that 'they' considered all of the right data and ideas, and this is the answer. Some bias is set as precedent. As more time goes by, those precedents are more easily accepted and adopted as solved problems, with less and less follow-on scrutiny. This isn't some deliberate attempt to ignore errors, but the reality of a large and complicated team, doing hard jobs, building a track record of success, and under pressure to keep achieving.

In all of these 'human element' related cases, something distracts us or lures us into decisions that lose alignment to the real-time morality, that add Links in the Error Chain. This effect can be so strong that the team will ignore data that conflicts with our decision-making.

Ignoring the Simple, Obvious Error

In 1998, during a Shuttle-rendezvous simulation, the sky was falling – not literally, although the spacecraft was failing in several different ways. It was still flyable, and we continued our approach to ISS. One of the failures was a leak in a rocket-fuel tank. This was something that couldn't wait since explosive and corrosive propellant could collect around the leak in the tail of the Shuttle.

We followed normal procedure and burned that propellant until the tank was empty. This particular tank fueled a hydraulic pump, not a normal rocket engine, so the exhaust was very small. It was small enough, in fact, that we didn't worry about the effect it could have on our orbit as we flew our 250,000 pounds mass Shuttle towards a 500,000 pounds mass ISS.

Over the next half-orbit, the astronauts called us several times, unable to find ISS when they looked for it out the window where it should be. That was puzzling, since all of the data in Mission Control told us Shuttle was right where they should be and closing fast. We had a separate display in the control room that showed us a different display from what the astronauts were using in the cockpit, and this was clearly showing a different answer from our display on the ground. It showed Shuttle way off the trajectory and nowhere close to being able to dock with ISS. The astronauts pointed that out to us.

That was clearly garbage data. We had two different sensors, sophisticated maths modelling (rocket science) and a crack team of experts producing our trajectory analysis in Mission Control; the astronauts had a laptop.

We gave them further direction for burns to clean up the discrepancy in the approach so they could see ISS. Still nothing. Still puzzling.

After another half-orbit, we gave up. ISS should have been so close it filled every Shuttle window, and still the astronauts couldn't see it anywhere. We clearly didn't know where the hell we were in the sky. With that realization, we stopped the simulation and went right into the debrief to discuss what we had just screwed up.

The Simulation Supervisor started the debriefing by asking me if I knew how much thrust the 'small' exhaust produced when we burned that leaking propellant. Turns out that my assumption that it was zero or negligible was wrong. It was small, all right, but large enough to affect the orbit because we hadn't turned the Shuttle to burn it in a benign direction, *which we could have.*

The reason Mission Control's data continued to look normal was that our calculated predictions were dominating the trajectory data we were projecting on the large screen in the control room. Unfortunately, the calculations did not include that 'small' exhaust, so our display was as ignorant of the real world as we were. We continued to be misled by that display since it showed us what we expected to see.

The astronauts' laptop was showing the real trajectory, just as other displays in Mission Control would have, had we looked at them. Those displays were continuously updating their trajectory information based on sensor data, not just the predictions. That meant that they weren't just showing our predictions and that they were correctly showing Shuttle flying further and further off course. We ignored it when we saw it on the astronauts' laptop because, 'It had to be wrong. We know where we are. Look, it's right there on our big screen.'

Great lesson to learn in training, although I kicked myself for a long time over that failure. Almost 20 years later, I still hang my head when I think about it. In flight, it would have cost us the mission. On a really bad day, we could have kept chasing the trajectory problem we didn't understand and collided with an ISS the astronauts couldn't see.

Besides the big, obvious risks we manage, like not blowing up the rocket, we have to manage the more subtle risk that is introduced through bias, subjectivity and errors that creep in, even unintentionally, from this wide range of 'human element' sources and then clouds our understanding and judgement. Once accepted, these biases and errors can go unchallenged despite new data or study that otherwise could have led the team towards truth. Further, once accepted by 'experts', heroes and respected leaders, these biases and errors become much more difficult to overcome. The world's best rocket scientists and most sophisticated analytics can't protect Mission Control from errors stemming from individual biases, preconceived notions, heartfelt but unexplainable certainty, or any other answer than the truth.

The real-time morality helps us there too, as it challenges the team and compels them to continuously ask the 'whys' and be on guard for the integrity in their thinking:

- Seek alternative perspectives, question your own objectivity and remain suspicious, if not downright paranoid, of biases – intentional or not – yours and the team's.
- Why do you agree? Why not?
- What if we were wrong before? What if we are wrong now?
- Are you sure you have enough information to support your conclusions?
- Are you open to new information that changes your previous understanding?
- What part of our decision-making is based on data or defensible rationale, and how much is unchallenged for one of the reasons above?
- What additional information or discussion could help reduce the uncertainty?

▶ The real-time morality challenges the team and compels them to continuously ask the 'whys' and be on guard for the integrity in their thinking.

▸ The real-time morality is key to managing the big, obvious risks – like not blowing up the rocket – and the more subtle risks that are introduced through bias, subjectivity and errors that creep in from a wide range of 'human element' sources.

▸ The world's best rocket scientists and most sophisticated analytics can't protect you from errors stemming from individual biases, preconceived notions or heartfelt but unexplainable certainty in any other answer than the truth.

Fly the Way You Train

Based on the real-time morality, we routinely pass judgement on our own performance, our peers', the team's and our leaders'. The litmus is the same: Were we working to the real-time morality as best we could? We judge ourselves and each other on both being right and in getting there the right way.

Judgement of our best effort isn't just success or failure. It is deliberately upholding that integrity, pursuing the 'whys', challenging and strengthening our grasp of technical truth. It is having the courage to then speak up, in spite of what may be a terrible risk, intimidating cultural norms etc, and make the call. Throughout it all, never lose alignment to the core purpose to protect the astronauts.

It is the discipline to always do this that separates great rocket scientists, flight controllers and leaders in Mission Control from the rest. The mantra in Mission Control is, 'Train the way you fly. Fly the way you train.' Absolutely right, in all things, big and small, whether we are fighting to prevent an explosion, responding to less critical failures or replanning the mission, we remember and apply what we've learned. This can become habitual, and it does for those of us who grow up in Mission Control.

Train the way you fly. Fly the way you train. Unfailingly. It's what we are compelled to do. Ask 'Why?' Challenge what we think we know. Go get truth. Always.

The way we train is all about unyielding alignment to purpose and deliberately applying the Mission Control trust elements of technical truth, integrity and courage *in all decision-making*. That is how we expect our people to fly.

That is the real-time morality. It is the key to Mission Control's legacy of highly reliable, real-time decision-making.

▸ Train the way you fly. Fly the way you train. Unfailingly.

Part 1
Finding Faith
The Real-Time Morality of Spaceflight

KEY POINTS

▶ The real-time morality: unyielding alignment to purpose and deliberately applying the Mission Control trust elements of technical truth, integrity and courage in all decision-making.

  ▷ Judgement of our best effort isn't just success or failure.

  ▷ It is all about relentlessly pursuing the 'whys', challenging and strengthening our grasp of technical truth, and having the courage to speak up in all decision-making.

  ▷ The real-time morality is the key to Mission Control's legacy of highly reliable, real-time decision-making.

▶ It is a morality – there are real consequences for every action we take, from financially costly to catastrophic and life threatening.

  ▷ The emotional charge that comes with that awareness strengthens the alignment to our core purpose and helps put the steel in the 'steely-eyed missile men and women' who have to make the calls.

  ▷ The emotional charge can also make the difference in tough situations to retain the discipline to take *deliberate, well-thought-out* action, instead of panicking and not managing the risk, not basing our decision-making on technical truth.

▶ The real-time morality is key to managing the big, obvious risks – like not blowing up the rocket – and the more subtle risks that are introduced through bias, subjectivity and errors that creep in from a wide range of 'human element' sources.

  ▷ Once accepted by 'experts', heroes and respected leaders, these biases and errors become much more difficult to overcome.

  ▷ As we achieve successful outcomes in spite of them, errors can be reinforced as correct decisions through the confidence we gain in 'success'.

▷ Through this confidence, we can learn to rely more and more on our judgement and intuition at the expense of deliberate thought *and technical truth, and it becomes progressively easier with each successful outcome.*

▷ This can also lead us to ignore new information that conflicts with our judgement and experience but could have led the team towards truth.

▶ The world's best rocket scientists and most sophisticated analytics can't protect you from errors stemming from individual biases, preconceived notions or heartfelt but unexplainable certainty in any other answer than the truth.

▶ Train the way you fly. Fly the way you train. Unfailingly.

PART 2

....

Losing Faith

Management Clouds the Morality

Don't say things. What you are stands over you the while, and thunders so that I cannot hear what you say to the contrary.

Ralph Waldo Emerson, Letters and Social Aims, 1875[18]

CHAPTER 4

Intentions and Practices

Many of us have forgotten what a real space flight operation requires. In the course of what seems to be endless meetings, simulations, schematic preparation, etc., it seems as though the actual flight will always be in the future. As a result, we tend to get more casual about our responsibilities and tolerate performance in ourselves and others that we would never accept in a 'flight mode' environment.

Pete Frank, Flight Control Division Chief, 1980[19]

Growing into an Enterprise

The business of human spaceflight grew up in a hurry. By design, as NASA incrementally learned how to fly and work in space on its way to the Moon in the 1960s, the technical challenges became progressively more difficult. The engineering effort required to meet those challenges kept pace in development, testing and manufacturing.

Setting aside the technical challenges and achievements, consider the investment required to support human space programs as a simple indicator of the size of the evolving enterprise. As the table below shows, whether in total cost or as a cost per flight, the growth from Mercury to Gemini is large, but from Mercury to Apollo is staggering. Total cost of the program increased by 70 times and the cost per flight by almost 40 times. Driven by Apollo's cost, NASA's total budget climbed from 1 per cent of the total federal budget in 1962 to 5 per cent in 1967 in preparation for the first Moon landing in 1969. As Apollo wound down and NASA began looking to new objectives and programs, its budget ramped back down to 1 per cent of federal outlays by 1974.

NASA Human Space Program Costs				
Program	Span	Total Cost $bn	Total Cost in 2010 $bn	Cost per Flight in 2010 $bn
Mercury	1959–63	0.277	1.6	0.265
Gemini	1962–67	1.3	7.3	0.723
Apollo	1959–73	20.4	109	9.9
Skylab	1966–74	2.2	10	3.3
	1975	0.245	1	1

(Claude Lafleur, 'Costs of US Piloted Programs', *The Space Review,* 8 March 2010)[20]

The investment included the large infrastructure required for the full range of technology development and testing, as well as two new NASA space centers, Johnson and Kennedy, which were responsible for spacecraft development and operations, and launch-vehicle assembly and operations, respectively. As the engineering efforts and infrastructure grew, NASA's management structure and process had to keep pace in order to maintain focus and control. This is summarized in NASA's 'Project Apollo: A Retrospective Analysis'[21]:

James E. Webb, the NASA Administrator at the height of the program between 1961 and 1968, always contended that Apollo was much more a management exercise than anything else, and that the technological challenge, while sophisticated and impressive, was largely within grasp at the time of the 1961 decision. More difficult was ensuring that those technological skills were properly managed and used.

Webb's contention was confirmed in spades by the success of Apollo. NASA leaders had to acquire and organize unprecedented resources to accomplish the task at hand. From both a political and techno-logical perspective, management was critical.

Reinforced in great part by the successes of Apollo, the human space program management structure and processes remained essentially the same through the next three decades. The total NASA budget hovered around 1 per cent of federal outlays through 1992 as NASA transitioned from Apollo to Shuttle and then ISS. That total ramped down to approximately 0.5 per cent, at which it has remained since 2006, although the total in dollars has increased with the federal budget. In its last few years of flight operations before retirement, the Shuttle program's typical annual budget averaged $4–5 billion. In that same time frame of 2004–10, the ISS program averaged $1.5–2 billion annually.

As NASA's human spaceflight effort evolved into a large enter-prise, so too did Mission Control. More specifically, the Mission Operations Directorate (MOD) – to which Mission Control belongs – grew as our full range of mission planning, training and flight control became more complicated. The original 20 scientists and engineers in a blockhouse at Cape Canaveral expanded rapidly with NASA in order to keep up with the dramatic increases in rocket, spacecraft and mission complexity.

By the early 1990s, the MOD total annual budget had grown to approximately $650 million. That budget remained relatively constant for the next 20 years, representing 6–9 per cent of each program we supported. Within that budget was a workforce that peaked at almost 5,500 people in 1992 while flying Shuttle and beginning preparation for International Space Station construction and flight operations.

Half of the MOD budget and workforce was dedicated primarily to the wide range of mission planning, training and flight control. Of this, most of the effort was in mission planning and training. As essential as flight control is, meticulous preparations have always been critical to Mission Control's success, as is the technical review process that ensures we've dotted all the i's and crossed all the t's before MOD commits to being ready for each next mission. Therefore, most of the effort and expense is spent planning and training for real-time operations, not in the flight operations themselves.

The Mission Control Center was more than doubled in size in 1993 to house the new space station flight controllers and support engineers for 20–30 years of around-the-clock operations, in parallel with ongoing Shuttle missions. Training evolved into full-purpose facilities such as the simulator complex, full-scale mock-ups and the Neutral Buoyancy Laboratory, with full-sized Shuttle and space station mock-ups submerged in an enormous 6.2 million-gallon pool for spacewalk development and training. The work required just to maintain these major facilities consumed the other half of MOD's annual budget and workforce for mission-related systems such as: dedicated data, voice and video networks; enterprise-computing and data-storage systems; simulator and control center software; water management and breathing systems; and more.

Like NASA, the MOD management team and processes matured and grew to keep pace with the MOD enterprise, although the advent of modern computers and network technology led to a steady reduction in the size of the MOD workforce from its peak in 1992. Even with those efficiencies, the workload associated with simultaneous Shuttle and ISS operations resulted in an MOD of more than 3,000 people. The resulting MOD management structure

Neutral Buoyancy Laboratory [NASA]

Space vehicle mock-up facility, housing full-scale mock-ups for
astronaut and flight controller training [NASA]

consisted of four levels of management, from first-line supervisor to the executive level of the Director of Mission Operations. By 2004, the organization was divided into 11 major divisions and offices.

Over the decades, there have been occasional MOD reorganization activities that moved boxes and lines around. The reorganizations moved functions from division to division, renamed divisions, and sometimes merged or split divisions. In each case, the general distribution of responsibilities remained the same and always aligned with our core functions. In the MOD management ranks, we came to see our roles as one of three functions:

- Plan/train/fly, meaning mission planning, training and flight control (MOD's reason to exist).
- Mission systems sustainment, meaning ongoing support and evolution of all computing and communications systems required by our simulators and the Mission Control Center (maintaining the most critical tools necessary to do the plan/train/fly job).
- Administrative and management integration (necessary to oversee human resources, financial performance and other functions to manage the enterprise).

With such a wide variety of technical expertise required, the plan/train/fly work is spread out across most of the divisions, typically organized by engineering discipline or similar technical focus. The plan/train/fly divisions and their subordinate branches are shown in grey in the figure on page 91. The unshaded organizations below fall into the other two categories, either mission systems sustainment or management functions.

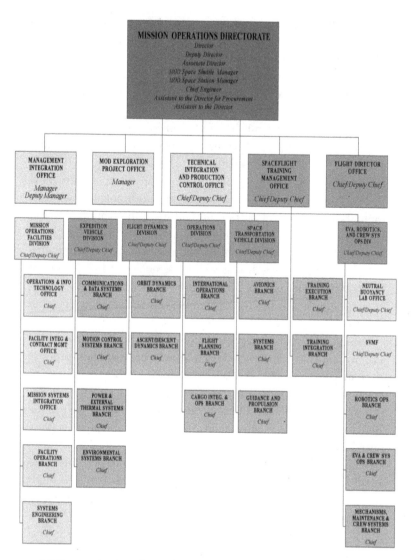

MOD organization chart, plan/train/fly organizations shaded (2011)[22]

Every flight control team is comprised of individual flight controllers, spanning the technical disciplines across all of the plan/train/fly divisions. Thus, the highest priority work MOD performs – real-time mission control – is performed by a matrixed team of working-level employees under the technical direction of a Flight Director.

The first level of managers focus on personnel development, mission preparations, project management, financial performance and similar standard management functions. The practice that has been followed since the 1960s is to promote strong performers at the working level into first-line supervisory positions. The real-time morality is a prevalent part of our culture at the working level in the divisions that support plan/train/fly work. Therefore, it is natural that the judgement of 'strong performance' is dominated both by technical expertise as well as demonstrated alignment to the real-time morality. So, of course, we look to the 'A-team' to supervise and train the rest of the rocket scientists and to mould them in their image.

The next level of management up, the first boxes on the organization chart, are then selected from supervisors who show promise as managers. Likewise, the Division Chiefs in the second tier of boxes on the organization chart are selected from those senior managers. The staff positions at the executive level as well as the Director and Deputy Director of Mission Operations are experienced leaders in MOD's critical technical work.

The beauty in our all-but-exclusive tendency to promote from within is that managers at all levels are steeped in 'the MOD way', the most important of which is the real-time morality with its strong alignment to purpose and emphasis on strong technical performance. We don't have to convince or convert managers as they're promoted. Since most come from a plan/train/fly background, they've already learned from personal experience how critical our role is and the importance of our highly reliable mission planning and real-time decision-making. They are accustomed to hearing (and already buy in to) the notion that, 'We're Mission Control, the conscience of human spaceflight'.

This alignment in our management ranks didn't just happen by accident. Everything from the vernacular we use in meetings and on the voice loops, to the focus and ideas that evolved into the real-time morality, reflect Gene Kranz's influence. That influence started early in his tenure as one of Chris Kraft's most trusted lieutenants, becoming more pronounced when he was selected to be one of the first four Flight Directors. By 1974, Gene was responsible for all plan/train/fly work. He was promoted to Director of Mission Operations in 1983, from which he retired in 1994.

For more than 30 years, Gene was the charismatic voice and vision behind MOD's leadership focus and in preserving a wide range of 'tough and competent' talk as the norm for MOD's rocket scientists. With our common 'upbringing' and Gene's strong presence, it's no wonder that the MOD management team has always had such strong alignment to our role and the ideas embedded in the real-time morality, especially in real-time in the Mission Control Center.

As managers, our time is consumed by management responsibilities (personnel actions, paperwork, budgets etc). It is the nature of the job, after all, when we move up from being rocket scientists and are charged with running the place. However, we tend to see it as being bogged down by less critical management stuff that separates us from the 'real work' that our organization prizes so highly. Over time, the question that dogs many of us as we are promoted up and out of Mission Control is how to apply the same values as managers, or even whether they apply in the management ranks at all.

▸ As NASA's human spaceflight effort evolved into a large enterprise, so too did MOD, growing into a $650 million per year effort with more than 3,000 employees in 2004.

▸ MOD's senior managers still professed the MOD 'faith' – that MOD was different, our responsibility was different, and that what our people did not only mattered but had the potential of 'ultimate consequences'.

- ▶ Our all-but-exclusive tendency to promote from within ensured that managers at all levels were steeped in 'the MOD way', particularly strong alignment to purpose and an emphasis on strong technical performance.
- ▶ Because our core purpose and fundamental role had not changed as we grew, MOD management prized mission-related work – both real-time and the preparations to get there and ensure success – but not much else.

Managing the Enterprise Presents Challenges

NASA's human space program grew significantly over the years. MOD followed suit and grew into a large enterprise in our own right, with a senior-management team that was three times the size of the entire flight operation's cadre from 1961. We then found ourselves struggling to demonstrate the values we had learned so strongly at the working level, and that led to management practices that were much different from the behaviours we demanded from the real-time team in Mission Control.

In 2004, 10 years after Gene Kranz retired and turned MOD over to a series of directors he had groomed, the management team still professed the MOD 'faith'. That is, it was still acknowledged in the senior-management ranks that MOD was different, our responsibility was different, and that what our people did not only mattered but had the potential of the 'ultimate consequences', as shown in The Foundations of Mission Operations. Because our core purpose and fundamental job had not changed, MOD still held our working level to ferociously high standards in real-time.

In practice, things looked much different in the management ranks than they did in the control room. Something had clearly been left behind. Outside the technical work, away from the control room and the mission – away from real-time – we had lost much of that single, unifying purpose.

The divisions had grown into stovepipes that operated as

independent fiefdoms. Except for mission-related technical items, key information (best practices, lessons learned, process escapes) was not communicated across division lines or up the chain to the executive level. Typical division chiefs' advice to their deputies and other subordinates who might attend a senior-management meeting was, 'Don't show your cards in front of the other divisions or the upper management. Fix your own problems, and manage your own work. All they're going to do is "help" anyway, and you don't want that kind of help.'

By 2004, behaviours that were in conflict with the real-time morality had become the norm among MOD's senior managers – a norm that still would not have been tolerated from the working level in Mission Control. These same managers, who genuinely professed the MOD 'faith', shared no practical semblance of collaboration on the common goal of managing MOD and stewarding a culture that ultimately was responsible for Mission Control and the single purpose of protecting the astronauts and the mission.

In real-time, as we saw in Chapter 2, we judge the flight control team on more than just 'talking the talk'. We place great value on how they do the job, because the alignment and integrity have such a powerful effect on highly reliable decision-making. In a similar vein, let's look at the expectations and normal *practices* that defined MOD's senior-management environment in 2004.

As an operations organization, MOD management valued mission-related work, both real-time and the preparations to get there and ensure success. The senior-management focus was on overall safety of flight and managing the planning and preparations for each mission. Technical consistency in the rocket science was delegated to the Flight Director Office in final reviews with all senior managers before Flight Directors led Mission Control in conducting each next mission. Leading up to those reviews, several management panels comprised of senior leaders from across MOD managed our mission preparations and ensured we were ready on time, and within our normal risk posture.

The financial baseline and performance of the $650 million enterprise was managed above the divisions at the executive level, and by 2004 this was handled by the Director of Mission Operations personally. Although most of the work in the plan/train/fly divisions was similar, each division's management process and rationale for resources was different. Each year's budget exercise depended heavily on what each division had the previous year, with a goal to 'circle the wagons and keep what you got'.

Not only did the budgeting process vary by division, but there was also no formal baseline or configuration control. During the year, if a division found itself in need of additional resources, the Division Chief consulted with the Director in private. As in any large enterprise, things happen that sometimes yield surplus funds as the year goes on. Sometimes a job takes fewer hours, actual investments are lower than predicted, or labour rates turn in our favour, all of which progressively accrue a surplus. Since financial performance was managed at the top, only the executive level had insight into this growing surplus and would dole it out when necessary. It was all very ad hoc, and was all about 'being first to the trough while it lasted'. It was handled privately in order to avoid a rush of divisions requesting help that would eat up the surplus. Once the yearly budget was allocated, each division was left to spend it all as they saw fit, as long as they got the expected work done and didn't overrun the budget.

All along, an MOD business office existed whose ostensible purpose was managing the financial baseline. By 2004, this office served largely as the liaison to our customers' and the space center's business offices to coordinate top-level accounting and resource transfers, in addition to specific analyses and reports for the Director, who was orchestrating the 'circling of MOD's wagons' himself. The MOD business office had no authority to manage financial performance with any of the divisions, since that was handled by the Director.

Of course, running a $650 million per year enterprise entails some regular management engagement and flow of information.

All senior managers participated in a teleconference at the start of each day to review the status of facilities overnight, mission preparations for the next mission, and issues with the day's training and flight operations. There was a variety of management-status meetings throughout the week, and each division submitted weekly activity reports.

Beyond these formal management processes, there were specific proprieties that had become accepted normal behaviour and expectations from the Director. Chief among these was the expectation that no one would take actions that would generate concern and attention from outside MOD.

MOD's image of being in tight control and needing no help or attention from our stakeholders was fiercely guarded. The last thing we needed was any senior executive at the Johnson Space Center or NASA Headquarters getting the impression that MOD had slipped in any way. This could only bring trouble in the form of losing some of 'what we got', if not outright meddling in how we did our rocket science from outsiders who didn't share the MOD 'faith'.

Thus, well before 2004, the overall guidance in MOD's senior management was, 'No ripples in the pond'. If a division could manage that, they were left alone to manage as they saw fit. They were 'trusted to just do the right thing'. If a division had technical or resource issues, they were expected to be discussed and resolved between the division and the executive level in private. That way, the Director could do damage control, where the first damage in question was any amount of concern or ripples in the pond.

This was no idle concern. The response to violating this guidance could be severe and public. A Division Chief who fessed up in a public meeting that their division was overrunning some work, needed some engineering help, or God help them, *might not be ready to fly on time*, could expect to be chastised in no uncertain terms. The expectation from the Director was that you were going to make this problem go away. Just the rumour of a problem or some sign of MOD weakness was considered a ripple in the pond.

As one Director said to a Division Chief, 'Why are you telling me about this? If I have to help you, what the hell do I need you for? Do I need a new Division Chief?'

This expectation to keep ripples out of the pond extended to all work, not just mission-related or rocket science. Concerns with personnel decisions, promotions, development project schedules etc were all worked out privately between the individual divisions and the executive level. In more public meetings, if a Division Chief disagreed with a peer organization's recommendations, they typically kept it to themselves. Doing otherwise was seen as 'throwing stones' and risked putting ripples in the pond.

This emphasis on keeping ripples out of the pond had other effects in turn. Something as straightforward as recognizing top performers was handled by dispersing awards evenly across all divisions on a per capita basis. When there weren't enough awards to go round, they were rotated based on 'whose turn it is this time'. The priority was ensuring no division was portrayed as favoured by their employees accumulating a disproportionate number of awards.

Since the divisions were expected to 'spend what they got' and there wasn't much collaboration, it became normal for multiple divisions to spend money developing their own division-unique solution for some common need. For example, there were at least three different software packages that were bought and customized to present the same data to flight controllers from different divisions who worked on the same flight-control team. This duplication of effort and investment ran the gamut, including flight controller software, spacecraft simulations, training-record management, and more.

Some of the divisions were able to subsidize their pet projects via secondary revenue they brought in by doing work for other NASA organizations. Much of this work was done without full insight from the executive level, but with the understanding that the division would get their other work done and stay on budget. However, once delivered, these pet projects and the proliferation of division-unique solutions typically increased each division's long-term fixed costs in order to maintain their duplicate systems. Those permanent increases were doled out from the main MOD budget a little at a time in private session with the Director. Since MOD's budget after 1993 was either flat or declining, these fixed-cost increases came at the expense of other work (similar to the non-discretionary budget items dwarfing the discretionary items in the federal government's budget).

The tables below list the basic senior-management governance and financial practices, and what evolved from them as the MOD-management cultural norms:

Basic Governance
Focus on safety-of-flight issues, flight readiness and flight control, and control team technical performance.
Technical consistency was provided via the Shuttle flight-preparation processes and flight rule control by the Flight Director Office.
Directorate-wide panels were primarily limited to flight production/readiness, with little other cross-division collaboration.
The Director's direct reports participated in daily telecons, submitted weekly activity reports and participated in weekly management info meetings.
Technical and resource issues were typically resolved between a division and the directorate in private.
Recognition opportunities were dispersed across all divisions and offices, civil servants were given preferential assignments, and rewards for top performers were often limited to eventual promotion.
Personnel decisions, promotions etc were worked out between individual divisions and the directorate office.
It was normal to have division-unique, duplicate solutions for common problems (flight controller certification metrics, workforce basis-of-estimates etc).

Financial Management
A financial baseline existed only at the directorate level, sometimes integrated by the Director personally.
No consistent division approach to cost Basis of Estimates.
Investments were largely 'first to the trough' in private discussion in the directorate office, governed by the available budget – spend what you got on what you see fit, as defined by each division.
No baseline-configuration control.
Primary resources book kept by the directorate, with margin sometimes parsed out, sometimes used as a 'repository' for a customer. Some secondary resources worked at the division level with no insight from the directorate.
A directorate office existed for center business office interaction with little formal management process and authority.

Directorate Level Norms (focused on conflict avoidance)
Best practices were not shared across division lines or with the directorate.
Diplomacy over clarity: Reveal as little as possible of any challenge to the directorate and other divisions, fix your own problems or risk getting 'help'.
Divisions were concerned primarily with their own area of responsibility and success, not necessarily with the directorate's – most divisions saw themselves in competition with the other divisions, rather than working towards common MOD goals.
Divisions were expected to take no actions that could generate concern outside MOD.
'No ripples in the pond': Divisions were expected to 'do the right thing' but take no actions that could generate concern outside MOD.
Don't 'throw stones' at other divisions' technical performance and personnel; just manage yours as you see fit.

After decades of achievement, MOD had grown significantly in size, budget and management complexity, while keeping pace with our mission complexity and similar growth across NASA. Along the way, our management practices reflected less and less the values and behaviours we still demanded from our workforce closest to the most critical risks we managed every day.

With the evolution to these management norms and practices, is it any wonder that by 2004 the divisions were concerned primarily with their own area of responsibility and success, not necessarily with MOD's? These senior leaders' organizations managed the enterprise, maintained the critical systems, and did all of the plan/train/fly work that MOD existed for, yet they saw themselves as competing business units. The divisions had more immediate problems than working towards common MOD goals or stewarding a culture: No ripples in the pond; protect what you got; keep the cards 'close to the vest', away from your peers and the boss.

As we would come to describe it in years to come, the value dominating MOD's senior-management practices by 2004 was diplomacy over clarity.

And it was *intentional*.

▸ As an organization grows into a larger enterprise, the financial and people-management challenges naturally grow too.
  ▷ The other costs and management responsibilities required to conduct business consume a growing percentage of the managers' attention versus the actual core business or product line.
  ▷ A normal tendency to focus exclusively on the core business or product can lead a management team to undervalue the other necessary management functions, and this can lead to practices that are inconsistent with the organization's best interests and business.
  ▷ Over time the inconsistent management practices can take on a more intentional focus than a management team's alignment and collaboration in their core business, which further erodes management performance.

- In Mission Control's case:
  - By 2004 there was no practical semblance of collaboration on the common goal of managing MOD and stewarding a culture that ultimately was responsible for the core purpose of protecting the astronauts and the mission.
  - The senior managers saw themselves as running independent and competing business units.
  - Management behaviours that were in conflict with the real-time morality had become the norm – a norm that would not have been tolerated from the working level in Mission Control.
  - The value dominating MOD's senior-management practices by 2004 was diplomacy over clarity, as described by: No ripples in the pond; protect what you got; and keep the cards 'close to the vest', away from your peers and the boss.

CHAPTER 5

Applying the Morality as Managers

We were a ruthless bunch in the control center. After every simulation, I demanded that each controller critique his own performance and the performance of his supporting staff. I'd add my own critique, and so would the simulation supervisor. It was done in public, all of us together, and before long we'd all developed a level of trust and camaraderie that I'd never seen before.

Chris Kraft, Flight: My Life in Mission Control[23]

Passing Judgement

Let's now turn the same ruthless eye that MOD has always focused on our real-time performance in Mission Control to our management team. As we learned in Part 1, judgement of our best effort in Mission Control isn't just success or failure. In the control room, we expect *deliberate* decision-making while under pressure and actions aligned to our core purpose. We judge ourselves and each other on both being right and in getting there the right way. That means relentlessly upholding the integrity, pursuing the 'whys', and challenging and strengthening our grasp of technical truth. It is having the courage to then speak up, in spite of what may be terrible risk, cultural norms and other intimidations.

Apply that judgement against MOD's management practices in 2004: no ripples in the pond; protect what you got; and keep the cards 'close to the vest', away from your peers and the boss. To do that, imagine a flight control team that conducts missions with MOD's management practices replacing the real-time morality. Managers would shy away from difficult decisions and speaking up in deference to keeping ripples out of the pond. In place of Mission Control integrity – that series of 'whys', ensuring our decision-making is deliberate and aligned to our core purpose – we'd focus on standing pat, not rocking the boat. Further, individual flight controllers keeping their cards 'close to the vest' while flying the spacecraft would cost the entire team technical truth.

All of these behaviours in Mission Control would come at the expense of the real-time morality. A Flight Director's judgement of any single flight controller conducting themselves like this would be immediate and harsh. Too much is at stake: the astronauts, the spacecraft and the mission.

The MOD senior management's judgement of a flight control team that was guilty of these behaviours would have been equally harsh. The initial motivation would be the same as the Flight Director's, because these behaviours would be a breach of our core values – the real-time morality that we all had imprinted into our DNA before we were managers. Unfortunately, part of the motivation for this judgement at the most senior levels would also be that if Mission Control had made some mistake, it would put some very large ripples in the pond. Ripples draw attention and help we may not want.

This is the management environment that Allen Flynt found when he took charge as the Director of Mission Operations in early 2004. He had been warned of these issues by the Deputy Director of the Johnson Space Center, Randy Stone, who had shared his concerns privately about MOD's management. This was important because, in addition to being Allen's new boss, Randy had been Director of Mission Operations from 1997 to 2001. Although confidence was still high in our technical work, MOD was seen by higher management as a 'stagnant, marching army' that had perfected circling the wagons and resisting all pressure to innovate and reduce costs.

It was those concerns that led Randy to seek out Allen (who had never been in MOD at any level) to fix things as MOD's new Director. By bringing in Allen, Randy was bypassing the entire MOD senior-management team, signalling that, for whatever reason, MOD's great rocket scientists were not turning out to be great senior managers.

Having watched and been impressed by MOD from the outside throughout his career, Allen went into his new role expecting to solve some communications gaps in an otherwise powerful and capable management team. This is Mission Control, after all. These guys get it done.

His first inkling that things might be a little more challenging than he had imagined came just before the first meeting with his senior managers on his first day on the job. His new deputy pulled him aside to prepare him for the meeting and ensure that Allen understood the top priorities. As he drew a map of the conference room, he told Allen, 'Listen, guy – it goes like this. You sit here, at the head of the table, and I sit on this side. Heflin (Milt Heflin, Chief Flight Director) talks first and sits on the other side. Then Jack, Epps, Larry . . .' His deputy's interest was not about the top list of rocket science, personnel or business issues facing the team; it was not about our most important development projects; and it was not a summary of the strongest and weakest members of the management team. As his deputy went through the entire roster of division chiefs and key staff members, Allen realized that the critical priority in mind was pecking order and not slighting anyone by getting that propriety wrong.

With that inauspicious start, and Randy Stone's warning, Allen quickly saw sign after sign of the emphasis on not putting ripples in the pond, protecting what you got, and keeping the cards 'close to the vest'. While the specific management practices described in Chapter 4 bothered him, his chief concern was more fundamental: the organization-wide stovepipes, low transparency, low engagement and entrenched senior managers.

This wasn't Allen's first executive position. He had been well

groomed through a steady series of progressively higher management responsibilities. He was both a natural leader and an
experienced senior manager. As such, Allen took a logical next
step to bring his new team together. In November 2004, he
conducted an offsite retreat for the entire leadership team shown
in the organization chart in Chapter 4. His stated goals were
unsurprising for any new leader of an established organization:

- Decide where MOD needs to be headed and what we need to
  do to get there.
- Identify MOD strengths, weaknesses, threats and opportunities (SWOTs) for each major scenario.
- Identify high-impact actions MOD can initiate; continue to
  leverage strengths and opportunities and mitigate or manage
  weaknesses and threats.
- Determine accountabilities for actions, general communication
  plan, initial and next steps.

Allen's real goal was to rally MOD's accomplished senior
managers around these discussions to bring them together as an
aligned team. That aligned team could then do something about
the stovepipes and wagon-circling in our management practices.
To hedge his bet, since he was still an 'outsider' to MOD, Allen
brought in an organizational-dynamics professional, Deb Duarte.
His hope was that Deb could get the management team to engage
on these subjects.

After two days of talking through those topics in several smaller
groups, the results weren't encouraging:

- MOD did not have an integrated strategic plan that defined
  operations concepts, staffing profiles and facility resources
  required for our evolving customer needs.
- The new human space program was not consistently asking
  for (or accepting) MOD input on spacecraft design and
  mission planning.

- MOD was not proactively working to give quality inputs to the new program.
- MOD was in need of strategic partnerships to ensure we retained responsibility for operations in the new program.

There was an ominous significance in our lack of strategic planning and our poor relationship with the new program. This new program was developing new rockets and spaceships that would replace Shuttle, take astronauts back to the Moon, and eventually take us to Mars. And it was this next human space program's managers who were not convinced they needed or wanted MOD's participation. This was a key part of the concerns that led Randy Stone to seek out Allen in the first place.

Allen assigned all of the senior managers to teams that were specifically focused on one of the four results from the discussion groups. They all then went back to work managing MOD. Over the next year, Allen continued pushing the MOD senior management to open up, talk about the team's weaknesses, implement changes and restore confidence from our stakeholders.

Meanwhile, the teams that had been formed during the 2004 retreat made no progress, aside from confirming what they already knew: we were doing pretty well at what we were tasked to do, as demonstrated by our success in flight. MOD was a very large organization, with a very important role, and we shouldn't take any actions that could upset the balance (and put ripples in the pond).

Allen closed out the year with another senior-management retreat in November 2005. Hearing the ongoing resistance to changing MOD's management practices, his revised goal for the 2005 retreat was, 'Without changing "Who We Are", could we be more efficient in "How We Do It"?'

Clever. He recognized that we're MOD-proud and that those attitudes were reinforced by our long and successful track record. He also brought Deb Duarte back to facilitate non-threatening discussions about opportunities to reduce costs without dulling our technical edge.

Instead of answers, the MOD senior-management team walked away from the retreat with a new set of teams to study five different strategic areas in the next year. It was not lost on Allen that more than a year and two management retreats into his stint as the Director of this high-profile organization, all he had to show for it was yet another plan to keep studying. He got up to summarize the retreat's results in closing remarks to the full senior-management team. Before he started, Allen stood at the front of the large conference room, staring at the floor, thinking, 'What do we do? We can't make progress on any strategic issues. We aren't trusted by our stakeholders.' He looked up with a helpless expression, and his voice cracked when he said, 'Guys, I can't do this without you, but I don't know what to do.'

At the end of this retreat, Allen's patience was gone but not his determination. He was set on a very clear goal to align the MOD management team on values, priorities and strategic goals, and get rid of the division fiefdoms and stovepipes. To make that clear, he sent the management team a note titled, 'Aligning Our Behaviors with Our Values',[24] opening with:

> Over the last 18 months as MOD Director, I've been asked on several occasions and venues to speak about what I believe are the keys to being an effective leader. At nearly all these opportunities I've said that living your values is one of the most powerful tools available to you to help you lead and influence others. Values drive behavior. And our behavior paints a picture of what we stand for.

> The same can be said of effective organizations. MOD is no exception. To be effective in the current business climate, MOD needs a workforce whose individual behaviors are in sync with the values of MOD as well as NASA.

> And I've spent considerable time clarifying my own personal values. They are not just mine . . . they are rooted in MOD and NASA. I

hope to be modeling my values with my work here and in turn, I expect to see you enthusiastically acting them out as well.

Allen's note goes on to describe in detail the specific values of 'respect, honesty, clear communication, acting responsibly and within a context of accountability, trust and humility'. One passage gives great insight into his frustration in tackling the challenges with this management team:

> Putting aside unnecessary suspicion of our colleagues and co-workers regardless of our agreement or disagreement with an idea spoken during a meeting or made during a presentation is essential to building the climate of respect I believe is necessary to move us to greater mission success.

> Sometimes, it seems that truth-telling becomes an excuse for a fundamental distrust of colleagues.

Again, this environment and these behaviours would never be tolerated in the Mission Control Room. These same senior managers wouldn't have tolerated it from their people while they were doing our most critical work. Using that same litmus, we should be harsh in our judgement of the MOD senior-management environment, if for no other reason than that their practices were incompatible with how we were raised – *by them*. It was incompatible with the real-time morality.

▶ We would be harsh in our judgement of the MOD senior-management environment that was in place by 2004 if we apply the same values to which we hold Mission Control.
  ▷ 'Diplomacy over clarity' and 'no ripples in the pond' conflict with the real-time morality.
  ▷ These same senior managers would not have tolerated the resulting behaviours from their people in the Mission Control Room.

Higher Stakes

The emphasis on no ripples in the pond – diplomacy over clarity – led over time to an entrenched and stovepiped MOD senior-management team. After 18 months, the new boss was not only unable to make meaningful progress breaking through, but he also found the need to chastise his most senior leaders about 'unnecessary suspicion' and the need to behave in a manner consistent with MOD and NASA values. In short, our senior-management practices did not reflect the real-time morality.

So what? Even if we agree with that judgement about MOD's senior-management practices, many of us fail to see a direct connection between flying rockets and managing our business. I was presented with that exact conundrum in a conversation with the CEO of a large global energy company. This accomplished executive got right to the point:

> I've heard you talk several times now about culture. I can see how that is important in Mission Control. I just don't see how it matters in running my business.
>
> Here's how it works for the executives in my company: I give you your annual targets. At the end of the year, if you hit your numbers, you did good. If you didn't hit your numbers, you didn't do good. Do that more than a couple of years, and I'll have to replace you with someone who can.
>
> How you do it, the culture in your area of responsibility . . . I don't know how that matters, as long as you hit your numbers and help the company reach our strategic goals.

To answer his point, we are focused on MOD management practices, not on rocket scientists flying the spacecraft in Mission Control. It is about management decisions that are necessary to keep a $650 million enterprise functioning from the top. And we were hitting our numbers. In 2004 and 2005, not only had MOD again performed flawlessly in operating the International Space Station, but we had also been key participants and leaders in safely returning Shuttle to flying in our first launch since the Columbia accident.

We had been challenged with one budget reduction after another for 10 years, which had steadily whittled down the size of our workforce, as shown in the plot below. However, we had managed to meet each of those reductions with a strategy that allowed us to keep flying Shuttle and to add ISS flight operations with a workforce that was 40 per cent smaller than it had been in 1993. Better yet, we had recently won some budget increases and projected some growth as we prepared to retire Shuttle in favour of a Moon and Mars program.

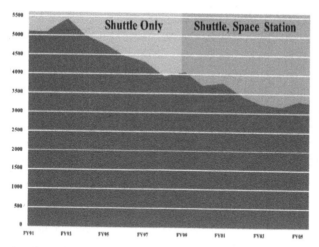

Number of people in MOD through 2006[25]

We were hitting our numbers, and we continued hitting our numbers. As we did so, in the background were Randy Stone's warnings to Allen Flynt about MOD's reputation in 2004, declining support for the organization as the operation organization of choice for any next human space program, and MOD management's inability to affect the outcome. Further, going into 2007, MOD had still made only small improvements in the senior-management environment.

In September 2007, as Allen's Deputy Director of Mission Operations, it was my job to sell the executives at NASA Headquarters on MOD's budget and contract strategy for the next five years. Our strategy was specifically intended to prepare us for lunar flight operations, and the overall plan was not surprising. It essentially reflected our budgeting and staffing experience of the last several years and projected them forward. We had come up with some innovations in the previous year that we thought would allow us to transition from Shuttle to the new program with fewer people in MOD and budget increases that only kept up with inflation. Buried within that plan was a several-year effort to modernize the Mission Control Center.

We felt like we had a good strategy since we had old, obsolete computers and networks in the Mission Control Center that we were nursing along until the end of the Shuttle Program. All of these needed to be replaced for the new program and the ongoing ISS operations. Rather than the hero's welcome I expected, one of the top executives pulled me aside to explain a growing concern at NASA Headquarters:

The MOD story is pretty good. You guys have done good bringing costs down. But it's still a lot of money.

And here's the thing. All of the centers have obsolete infrastructure that has accumulated through almost 30 years of Shuttle operations. All this old stuff takes a large workforce and budget to keep it working. And every center has a plan to keep all of their old stuff and all of their people. Then they have to add new stuff – new infrastructure, new people – new cost in order to deliver the new content associated with the new program, right? All of the new stuff just gets added to the existing base.

So by definition, the next program will cost exactly what Shuttle costs *plus* the cost for all of the new stuff. Rather than saving money by retiring Shuttle so we can go back to the Moon, in our flat budget we're guaranteeing that we still won't have any money for new rockets and spacecraft.

We either get a budget increase so we can keep all of the old stuff and buy the new stuff – which isn't going to happen – or we can never solve this problem, and we can't do any next program.

Even after that conversation, I didn't understand just how unsolvable this impasse was considered to be at NASA Headquarters. Nor did I realize how strongly so many executives at that level saw MOD as just another 'marching army' that is willing to do things only in the way we did them in the 1960s. We found out two years later.

In the Fall of 2009, President Obama's first year in office was wrapping up, and the administration was pressurizing NASA to eliminate all traces of the Bush Vision for Space Exploration. Compounding that pressure, Headquarters' concerns over the NASA centers protecting their 'old stuff' and driving up costs had continued to fester, and NASA leadership was ready to take dramatic action.

A few months later, in February 2010, NASA Administrator Charlie Bolden informed the public and the astonished NASA centers that the Moon and Mars program (Constellation) was being cancelled. Shuttle would still be retired in the next year as planned, and NASA would award contracts to private companies to fly our astronauts to ISS.

In a speech in Houston shortly after that announcement, Bolden didn't talk about the hundreds of millions of dollars per year that were encumbered in old infrastructure at all of the centers that were resisting innovation. Instead, he said, 'We can no longer afford to have thousands of flight controllers sitting over there in that windowless building flying our spacecraft.'

Not only did this decision cut MOD's budget and workforce in half, it also put us out of the space launch-and-landing business. When the last Shuttle flight landed the next year, we would lay off almost 90 per cent of the workforce who had any experience in launching and landing human spacecraft, as well as the analysis, mission planning and training those critical operations required. MOD, the conscience of human spaceflight, had been lined out of protecting our astronauts during the most dangerous phases of flight. And *we* were specifically cited as leading factors in the decision!

To add insult to injury, two months later I found myself in an informal gathering of 50 or so senior executives from NASA and industry. The room hushed as our host, an executive from one of the companies who was going to bid on NASA's launch-and-landing services, announced to the room that I was there on an interesting mission. Now the Director of Mission Operations, I was in discussion with a number of companies exploring the possibility of MOD providing services to them in support of NASA's 'commercial' contract. Before I could explain any merits in the strategy, the NASA ISS Program Manager, MOD's main customer, regaled the room with, 'Why the hell would anyone hire MOD? They're too expensive. They have too many people. All they know how to do is what they've always done.'

Our largest customer – the daily beneficiary of the 'Gods of Human Spaceflight' since ISS's first flight in 1998 – had just told leaders from all of our prospective new customers that they'd be fools to hire us to fly their spacecraft. *Holy shit!* My reaction at the time was much stronger than that, and very public. It didn't do anything at that moment to change the opinion of that program manager.

The irony is, in the previous two years, MOD had reduced the workforce required to support ISS flight operations by 23 per cent! Also, rather than thousands of flight controllers, a typical ISS flight control team was comprised of only 7–15 flight controllers, depending on the day's activities in space. On this day in 2010, the ISS Program Manager knew not only that we'd given him those savings but also that no other supporting organization had.

The reductions we had made in flying ISS weren't forced upon us as budget cuts from above. Instead, the internal MOD-management discussions that had frustrated Allen in 2004 and 2005 had finally started turning things around. By 2007 we had started making significant changes in MOD's management environment, which led directly to innovations not previously considered possible by MOD. MOD then turned those innovations into unexpected cost savings for customers. (We'll come back to the changes in the management environment in Part 3.)

Unfortunately, we hadn't had enough water under this bridge to change the mindsets of many influential stakeholders, including this program manager and the executives at NASA Headquarters. Since we were still seen through the old lens, and the agency-wide problem was so large, MOD was lumped in with everyone else as obsolete and ready for the scrap heap.

So, again, did it matter that our management practices hadn't reflected the real-time morality? Sure, we were still hitting our numbers, but over the years we had indelibly reinforced our image as a 'stagnant, marching army'. That reputation was so strong to many of our stakeholders that they didn't give MOD credit for innovations we had already made, like the 40 per cent reduction in the size of our workforce by 2004. As we discovered, they were no more inclined to give us similar credit now for more recent innovations. They seemed just fine with dropping MOD altogether and giving *anyone* else a try.

The NASA centers had just become too expensive and backward thinking, and MOD was seen as just another component of them. As one NASA Headquarters manager said to me while explaining their thinking, 'Look – we'll give this a try. Let the private industry try new ways to fly our human spacecraft and reduce costs. If it doesn't work, *what's the worst thing that can happen?* You guys invented MOD in the '60s. We'll just invent it again in 10 years if the experiment doesn't work.'

We discovered that our management practices did matter, even when we hit our numbers.

▶ Our management practices matter in other ways, even when we hit our numbers – the effects of diplomacy over clarity (no ripples in the pond) eroded the trust from our stakeholders and consequently cost us business opportunities.

At Our Core

Reputation and business-related impacts are one thing, but MOD's values are driven by the real-time morality not management mumbo jumbo and budget talk. Closer to home inside MOD, we worry about impacts to the rocket science, or more specifically, on the high standards we hold ourselves to in support of the real-time morality. As long as we are known to be on the right side of every rocket science-related issue, we'll take our chances on what the outside world thinks about our management style. However, no ripples in the pond and keeping cards 'close to the vest' was finding its way closer to our rocket science.

....

Management Practices Ripple Towards the Mission Control Room

By 2006, ISS construction in space was well under way, with more than 500,000 pounds of spacecraft already assembled, on the way to more than 1,000,000 pounds when completed. In November 2006, the MOD senior management convened a flight-readiness review for STS-116, the next flight of Space Shuttle Discovery to ISS. This mission included some complex robotics and spacewalking, and it laid the groundwork for even more complex robotic operations that would soon follow.

In MOD's flight-readiness review, the lead Flight Directors and every division reviewed every detail of the mission, our preparations, and any remaining issues that needed to be resolved before flight. The point was to not leave any stone unturned, sometimes requiring two days to discuss everything. For the conscience of human spaceflight, time isn't the issue – thoroughness is. This has been MOD's standard practice for decades.

As always, this was a good review of an extremely well planned and difficult mission. As the lead Flight Directors wrapped up their summary of MOD's overall readiness for flight, the ISS Flight Director made it clear: 'We are go for STS-116.'

As he did so, most of the MOD senior managers around the conference table were nodding their heads, expecting to wrap up the rest of the review as a formality and declare us ready to fly. At that time, I was MOD's Manager for Shuttle Operations, and I needed some clarification. 'Didn't you say that we had recently been given new loads constraints for the solar arrays? And the PHALCONs (ISS flight controllers who operate the electrical system) don't yet understand how to manage the solar arrays within those constraints? So we don't know how to safely fly ISS at the end of the mission?'

The ISS Flight Director said, 'For the purposes of STS-116, we're good to go.'

I asked, 'Let me make sure I understand. Your team has everything they need to manage the solar arrays *during* STS-116?'

'That's correct. We're go for the flight' he said again.

'Uh, huh. But the loads constraints change at the end of the flight?' I asked. The ISS Flight Director answered, 'Yes, after we've retracted the P6 port solar array, we'll have to rotate the port SARJ (solar alpha rotary joint) and track the sun to generate adequate power when Shuttle isn't docked. The new loads constraints apply when we start sun tracking, and we don't yet know how to meet them. But we're good for the flight.'

By now, most of the senior managers were looking puzzled. The PHALCONs' division chief leaned forward and surprised us with, 'My division is no-go for launch without this resolved.' Our boss looked down the table at me, and he was not happy.

'So let me be clear', I said. 'You're good to go *while* Shuttle is docked? *As soon as Shuttle undocks,* ISS will need to resume attitude control and rotate the SARJ to track the sun, and *we do not know how to do that* without breaking the space station?'

'Correct' was the Flight Director's answer.

I was baffled. 'Wait – how are we go for this mission? We know how to operate ISS *now*. We know how to operate it *during* the Shuttle mission. But because of what we do during the mission, we don't know how to operate the spacecraft *immediately after Shuttle undocks*?

'These aren't two separate events. When we launch STS-116 in one month, it's all or nothing. Not knowing how to safely operate the space station seconds after Shuttle undocks doesn't make us go for launch with some little technical detail to work out later. It means we're breaking the space station if we conduct the mission as planned.'

After some heated discussion, our boss made it clear. 'All right, guys, we are *no go* for launch. The ISS program will either have to do more analysis and give us relief on the loads constraints, or we're going to need an investment in engineering work to give us the method we don't yet have to manage the solar arrays within the new constraints. Until then, MOD is no go.'

In the next several weeks, we were given temporary relief of some of the loads constraints by the customer and launched STS-116 on time. The picture below shows ISS after Discovery had undocked. The solar array loads constraints remained a thorny issue and drove significant complexity in operating ISS for years to come.

....

Why did the PHALCONs' division chief wait until this review just before the launch to inform us they were no go? Because his people believed they had made this same point and had sought

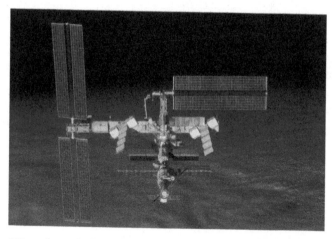

ISS at the end of a successful STS-116, 19 December 2006 [NASA]

help from the Flight Directors for months – and had consistently received the same answer: 'Don't worry, we'll figure it out'. His team hadn't raised the issue already because the flight-preparation process had largely been delegated to the Flight Directors, so the division's managers were reluctant to step outside the process.

The Flight Directors in question weren't newbies or afraid to take on tough problems. They were seasoned, brilliant and tough. They also knew this to be a contentious subject with our customer at a critical phase of the program, and they were trying to keep ripples out of the pond. They had good reason to believe the issue would be resolved, but it wasn't. Worse, the customer was not aware that we were in this position only a month before launch.

While listening to the discussion, the other MOD senior managers sat silently during our review because it wasn't *their* issue. This was between the Flight Directors and the customer (or in this case, between the Flight Directors and Paul Hill, who for some reason was stirring the pot).

Diplomacy over clarity was still with us.

We discovered that our management practices influenced more than budgets and other administrative work, after all.

This episode made many of the MOD senior managers very concerned that we could be dodging bullets, and that's not how we were raised in MOD. Our flight-readiness reviews are not intended to *catch* mistakes or issues – they are intended to review our status in *resolving* them.

This one worked out. We had caught and corrected the error in our judgement. But what if the next one was more critical and we didn't catch it? What if no other part of the community caught it either? In what other work were we already making similar decisions? What did this tell us about our management priorities when almost all of us not only missed this issue but were also not interested in taking ownership when it became apparent to us?

▸ Our management practices influence more than budgets and administrative work – in spite of our professed focus on always getting our rocket science right, 'no ripples in the pond' and keeping cards 'close to the vest' found their way closer to our rocket science.

No Higher Stakes

The previous examples still may not convince some managers of the direct connection between flying rockets and managing our business. From reputation and business impacts to the *possibility* of missing something in the rocket science, these aren't the same as actual failure. Besides, the risks that can lead to failure in flight are managed in real-time by Flight Directors, their flight control teams in Mission Control and the astronauts on the spacecraft. They have a tremendous track record of highly reliable decision-making and achieving mission success, and making it look easy.

Although NASA's higher-level strategy and MOD's have sometimes differed, we have always been aligned on our core purpose to prevent catastrophe – to protect the astronauts. Tragically, however, NASA has failed in that purpose three times (Apollo 1, Challenger and Columbia) with astronauts in harm's way in our spacecraft. In none of those cases did the real-time teams make mistakes that cost us those astronauts' lives. However, in each case, senior-management practices were key contributors, as the summaries that follow show.

Apollo 1

On 27 January 1967, the Apollo 1 astronauts were suited up and strapped into the Command/Service Module (CSM) at the top of the Saturn IB rocket for a live test in preparation for launch a month later. This was considered a non-hazardous test since the rocket wasn't fueled and all the pyrotechnics were disabled.

After sampling the air in response to an unknown sour smell in Gus Grissom's pressure suit, the hatch was bolted closed. The astronauts and the launch control team at KSC went through their checklists to configure the spacecraft just as they would during a launch. This included pressuring the cabin atmosphere with 100 per cent oxygen to 2 psi above the outside air pressure to provide positive pressure, like the spacecraft would experience in flight.

[top left] *Roger Chaffee, Ed White and Gus Grissom suiting up in the 'white room', 27 January 1967.* [top right] *Apollo 1 on pad 34, KSC 1967.* [bottom left] *Roger Chaffee, Ed White and Gus Grissom strapped into the Apollo 1 CSM simulator.*
Photos [NASA]

Five and a half hours into the test, a fire ignited inside the cabin near the floor below Gus Grissom's feet, where electrical wires were routed near the oxygen panel. Less than a minute later, the oxygen-fed fire had consumed the interior, ruptured the pressure hull of the CSM and killed all three astronauts.

Although no single ignition source of the fire was conclusively identified, the Accident Investigation Board identified four conditions that led to the fire:

- A sealed cabin, pressurized with an oxygen atmosphere.
- An extensive distribution of combustible materials in the cabin.
- Vulnerable wiring carrying spacecraft power.
- Vulnerable plumbing carrying a combustible and corrosive coolant.

As a result of the fire, NASA abandoned this version of the Command/Service Module. In addition to accelerating the Block II CSM, NASA made a long list of modifications to resolve safety concerns related to the CSM, the pressure suits and the communications systems at KSC. Of course, a key change to the CSM was switching to an oxygen–nitrogen cabin atmosphere to reduce the risk of fire.

With the fantastic successes in Mercury and Gemini, how had we come to this? Recall from Chapter 4 that the Apollo Program represented a 70 times larger investment than Mercury, with Apollo costing 40 times more per flight than Mercury. Those cost increases were a direct result of the increased complexity of Apollo's design, development, testing and manufacturing. This growth and complexity had consequently become a much larger management challenge.

Also, NASA was consumed with meeting John F. Kennedy's commitment to reach the Moon within the decade, now even more so to honour the legacy of the assassinated President. Increasing that pressure, not only was the United States not yet beating the

Soviet Union in human spaceflight achievement, we had been playing catch-up since Yuri Gagarin's first human spaceflight in April 1961 (if not since Sputnik in 1957).

One unfortunate side effect of the Cold War was that NASA was unaware of the Soviet Union's similar experience with their own oxygen-driven fire, although the Soviets' was in an altitude chamber, not a spacecraft. That fire cost the life of cosmonaut Valentin Bondarenko and prompted the Soviets to eliminate the 100 per cent oxygen atmosphere in their spacecraft. The Soviets had learned this hard lesson in 1961, *six years before* the Apollo 1 fire. NASA had no knowledge that it had even happened or that the Soviets had abandoned 100 per cent oxygen atmospheres.

Thus, by 1967, the Apollo effort had grown tremendously: NASA was under great pressure to reach the Moon on schedule and beat our Cold War adversaries; and Cold War politics prevented the exchange of information that very likely would have alerted NASA to this oversight in high-pressure, oxygen-rich environments.

Apollo 1 crew appealing to a higher power over problems with the Command/Service Module [NASA]

But before the fire, the Apollo community was aware of a wide range of engineering problems with the spacecraft, communications systems at KSC, and more. In spite of those known problems, Apollo 1 was scheduled to launch only a month after the ill-fated test. The Apollo 1 astronauts even sent the photo below to Apollo Program Manager Joe Shea, with the inscription, 'It isn't that we don't trust you, Joe, but this time we've decided to go over your head.'

However, there had been hurdles before, and NASA had always cleared them. Surely these would be resolved as well, and we must keep up the pace! This is exactly what Gene Kranz was referring to in his speech[26] to Mission Control after the fire when he said:

> We were too gung-ho about the schedule, and we locked out all of the problems we saw each day in our work. Every element of the program was in trouble, and so were we. The simulators were not working, Mission Control was behind in virtually every area, and the flight and test procedures changed daily. Nothing we did had any shelf life. Not one of us stood up and said, 'Dammit, stop!'

Challenger

Nineteen years later, on 28 January 1986, Space Shuttle Challenger launched on mission 51-L. Challenger was carrying two satellites to deploy in space, a number of experiments, a civilian teacher to stimulate education, and of course six more astronauts for a crew of seven.

Seventy-three seconds into flight, a jet of hot gas leaking through the seals between two lower segments of the right Solid Rocket Booster (SRB) burned through the external tank, igniting the liquid hydrogen inside. The external tank exploded, destroying Challenger. The still-intact crew compartment continued upward to 65,000 feet in the sky before falling. It impacted the Atlantic Ocean 2 minutes 45 seconds after the explosion, killing all seven astronauts.

[top left] 51-L astronauts going to the pad for launch. [right] Challenger at lift-off on 51-L, 28 January 1986; smoke emanating from the SRB in lower right. [bottom left] 51-L flightdeck crew in a cockpit mock-up during training.
Photos [NASA]

The SRB joint O-ring seals leaked because they were too cold when the rocket engines were ignited. The O-rings therefore did not respond as designed, allowing the burning rocket exhaust to blow past them. Once through, the hot gas eroded the seals further and fully burned through the final seal. The leak started almost the instant the SRBs were lit, as can be seen by the small cloud of smoke from the SRB in the lower right of the launch photo above.

Before 51-L's launch, there had been growing concern about O-ring performance at both the manufacturer, Morton-Thiokol, and by some engineers at the Marshall Space Flight Center, who managed the SRBs for NASA. Post-flight inspection found that SRB O-rings had experienced some amount of hot gas blow-by on 12 of the previous 15 Shuttle launches. Further, significant charring was found on the final seals of three SRBs used in the previous year.

Although they did not have data to prove exactly how low

temperature affected the seals, Morton-Thiokol engineers were convinced temperature was a significant contributor to this problem. The coldest test firing of an SRB had been at 50 degrees F. Because of the results from that, and other test firings and previous Shuttle launches, the company considered 53 degrees F to be the minimum safe temperature to fire an SRB.

Freezing weather in central Florida that January had created conditions for the coldest launch attempt yet. In meetings leading up to the launch, Morton-Thiokol engineers were adamant that the launch should be delayed because of temperature-related concerns with the SRB joint seals. Their senior managers were fully supportive, as were engineers at Marshall. NASA's SRB Project Manager at Marshall was unconvinced and pressured Morton-Thiokol to prove their concern, challenging them in one meeting, 'My God, Thiokol, when do you want me to launch – next April?'

In the last meeting before giving a final answer regarding the launch, Morton-Thiokol executives asked their Engineering Vice President to 'take off his engineering hat and put on his management hat'. As described in testimony to the Presidential Commission on the Space Shuttle Challenger Accident[27]:

> This was a meeting where the determination was to launch, and it was up to us to prove beyond a shadow of a doubt that it was not safe to do so. This is in total reverse to what the position usually is in a preflight conversation or a flight-readiness review. It is usually exactly opposite that.

After this meeting, Morton-Thiokol officially recommended proceeding with launch with no new data or engineering explanation. NASA's SRB Project Manager recommended proceeding with launch to the NASA executives at KSC with no mention of Morton-Thiokol's or other MSFC engineers' concerns with cold O-ring performance.

Ice on pad 34 before Challenger launch on 51-L, 28 January 1986
[Photos: Presidential Commission on the Space Shuttle Challenger
Accident, Report to the President]

The air temperature on the pad at launch time was 36 degrees F. The right SRB lower joint seal temperature was 28 (±5) degrees F at ignition, well below Morton-Thiokol's recommended minimum temperature of 53 degrees F.[28]

After learning a hard lesson with Apollo 1, the tremendous successes of Apollo, Skylab, Apollo-Soyuz and the previous 24 Shuttle flights, how did we come to this?

A new kind of schedule pressure had developed. From 1981 through 1985, NASA had increased the number of Shuttle flights from two to nine per year while trying to reach an operational goal that fluctuated between 12 and 24. Shuttle development and infrastructure costs had exceeded original estimates, and flying fewer than 12 per year left them too expensive per flight to justify NASA's investment. If it looked like 'we lied to Congress', we might lose funding, stop flying Shuttle, and not build a space station. All of that meant we were at risk of stopping human spaceflight completely, never going back to the Moon or Mars, or anywhere else.

Also, it had been almost two decades since Apollo 1. NASA had learned and accomplished many things in space in that time. The executives leading at Headquarters and at the NASA centers were predominantly experienced and respected leaders who had built careers leading those accomplishments.

Thus in 1986, as in 1967, NASA had big challenges. They were still managing a multi-billion dollar, nationwide, multi-center and multi-prime-contractor effort, and they were ramping up the effort to at least 12 flights per year. Confidence in their management ranks had grown with each success, although some of the management practices had not kept pace.

The Presidential Commission on the Space Shuttle Challenger Accident summed up their criticism of the contributing management practices[29]:

1. The Commission concluded that there was a serious flaw in the decision-making process leading up to the launch of flight 51-L. A well structured and managed system emphasizing safety would have flagged the rising doubts about the Solid Rocket Booster joint seal. *Had these matters been clearly stated and emphasized in the flight readiness process in terms reflecting the views of most of the Thiokol engineers and at least some of the Marshall engineers, it seems likely that the launch of 51-L might not have occurred when it did.* [Emphasis added]

2. The waiving of launch constraints appears to have been at the expense of flight safety. There was no system which made it imperative that launch constraints and waivers of launch constraints be considered by all levels of management.

3. The Commission is troubled by what appears to be a propensity of management at Marshall to contain potentially serious problems and to attempt to resolve them internally rather than communicate them forward. This tendency is altogether at odds with the need for Marshall to function as part of a system working toward successful flight missions, interfacing and communicating with the other parts of the system that work to the same end.

4. The Commission concluded that the Thiokol Management
   reversed its position and recommended the launch of 51-L, at
   the urging of Marshall and contrary to the views of its engineers in order to accommodate a major customer.

Columbia

Seventeen years later, on 16 January 2003, Space Shuttle Columbia
launched on mission STS-107. Columbia was carrying a SPACEHAB
Double Research Module, a number of experiments, and a crew
of seven astronauts. The following launch photo shows a debris
cloud after a piece of foam insulation called the bipod ramp sepa-
rated from the external tank and struck Columbia's left wing 81.9
seconds into flight. The astronauts were informed that evening
and told the Orbiter may have some tile damage that would need
repair after landing. On 1 February 2003, after a successful mission
in orbit, Columbia entered the atmosphere for a landing at KSC.
Because of damage in the leading edge of the wing, the fireball of
ionized air that normally flows around the Orbiter penetrated the
left wing, destroying it. Sixteen minutes before landing in Florida,
Columbia lost control and came apart almost 40 miles in the sky,
travelling more than 10,000 miles per hour, killing all seven astronauts.

As with all anomalies, the STS-112 foam loss was reviewed
before the next launch in November 2002. At that review, the
official explanation of the risk from foam debris fit on a single
PowerPoint chart. It said essentially that foam had been shed from
the tank for years and 'has never been a safety issue'. Hence, it
was acceptable to continue flying with this problem unresolved
– it was safe. The rationale was accepted, closing the anomaly.
That flight flew without a problem.

[top] *Columbia launching on STS-107, 16 January 2003.* [bottom left] *STS-107 astronauts going to the pad for launch.* [bottom right] *Pilot Willie McCool and Commander Rick Husband piloting Columbia during entry, 1 February 2003*
[Photos: Columbia Accident Investigation Board Report]

The next launch was Columbia on STS-107. Before launch, external tank foam loss was not discussed, because it had been declared a non-problem before the previous flight.

Overall, damage from the foam strike was not considered a critical issue during the flight. The community had already largely accepted foam debris as only a nuisance and had just been reminded of that bias after the STS-112 external tank foam loss.

The reinforced carbon-carbon on the Orbiter's wing leading edge was much less fragile than the easily chipped tiles, which frequently required post-flight repair. Therefore, the analytical focus during STS-107 was on potential *tile* damage, which was correctly determined to be acceptable.

The possibility that the leading edge was damaged was not only discounted by NASA during the flight – it was hotly contested by senior NASA managers as even a possibility when the Columbia Accident Investigation Board pursued it as a probable cause from the outset of the investigation.

Tests conducted after the accident showed conclusively that foam debris impacts duplicating the impact during Columbia's launch were capable of creating critical damage to the wing leading edge. Data reviewed from recorders recovered from Columbia's wreckage showed just as conclusively that the damage was in the left wing leading edge.

How did we come to this, yet again?

First, a new schedule pressure had developed, again. After the original Space Station Freedom program was morphed into the International Space Station, development challenges led NASA and the Russian Space Agency to park the first elements of the ISS in orbit in November 1998. They knew then that the next major elements of the space station would not be ready to fly for more than a year, but they launched anyway as a way of mitigating political pressures to get started. The total space station investment continued to build up from previous redesigns, the change to ISS and the various development-related launch delays. Thus, NASA was under pressure to start launching the space station missions, finish the construction effort and limit the total cost. This pressure was felt at all levels, from executives who were answering for any delays to the working level, who were doing their best to deliver, as NASA always had.

NASA had also spent nine years trying to implement a goal to be faster, better and cheaper. As NASA resumed ISS construction in 2000, the programs were being squeezed to pick up the pace, in the most complex series of space operations in history, and with fewer resources.

Although faster-better-cheaper is an understandable management goal if implemented carefully, by 2002 it had a different effect in NASA's human spaceflight programs in combination with the ISS Program schedule pressure. With an ever-growing cadre of managers with lengthy experience in spaceflight, it became progressively more normal to trust the judgement or intuition of the leaders who had 'been there and done that'. Most of these leaders had reputations for solving hard problems, and each next success reinforced the confidence in their 'engineering judgement'. As faster-better-cheaper reduced the ability to analyse and test a wider range of issues, the community and these experienced managers relied on their 'engineering judgement' even more, often referring to it explicitly as justification in support of technical arguments.

Engineers raising concerns without ironclad data to prove their concern were more frequently turned away. It became a tug-of-war of judgement (or opinion), and the boss always won. Thanks to faster-better-cheaper, we didn't have the resources to do more analysis and tests to prove every concern was valid. The engineer accepted defeat and moved on to the next issue.

Many times, as a technical discussion wrapped up, the engineers were told to reconsider their recommendations and present them again at a later meeting. In the next meeting, their recommendations were expected to reflect the 'engineering judgement' of the respected executive leading the meeting. Not only did they typically make those changes, it had become such a normal practice that it wasn't seen as misleading or wrong.

And that led us right back to the position Morton-Thiokol found themselves in before the Challenger accident – the need to prove something is not safe, rather than providing evidence that it is. Because we moved in this direction over time, the 'normalization' was so thorough that many issues were never brought forward from lower-level reviews. Even working-level engineers would elect not to raise some issues because, 'You know we're not going to get our day in court on this. We can't prove we're right, and we're not going to get the money to prove it.'

Thus, in 2003 NASA was still accomplishing difficult things with a multi-billion dollar, nationwide, multi-center and multi-prime-contractor effort. The key executives were recognized experts, and their judgement was unchallenged.

Rather than costs per flight, our schedule challenge before Columbia was driven by completing ISS construction on time and limiting its cost growth. This was compounded by the faster-better-cheaper pressure to achieve more, in less time, and with less investment. Over time, the can-do NASA workforce kept delivering, and many decisions were made with less and less technical rigour.

Once again, some of the management practices were getting in our way.

The Columbia Accident Investigation Board[30] summed it up in their criticism of the contributing management practices:

> Within NASA, the cultural impediments to safe and effective Shuttle operations are real and substantial, as documented extensively in this report.

> The Board's view is that cultural problems are unlikely to be corrected without top-level leadership. Such leadership will have to rid the system of practices and patterns that have been validated simply because they have been around so long.

> Examples include: the tendency to keep knowledge of problems contained within a Center or program; making technical decisions without in-depth, peer-reviewed technical analysis; and an unofficial hierarchy or caste system created by placing excessive power in one office. Such factors interfere with open communication, impede the sharing of lessons learned, cause duplication and unnecessary expenditure of resources, prompt resistance to external advice, and

create a burden for managers, among other undesirable outcomes. Collectively, these undesirable characteristics threaten safety.

Common Themes from Apollo 1, Challenger and Columbia

In each of these cases, NASA was managing large enterprises, with cumbersome government management practices and a wide range of technical risks, and was achieving success after success in the exceptionally unforgiving environment of space. The words of the Columbia Accident Investigation Board[31] ring true through each decade, and in the time of each of these accidents:

> NASA is a federal agency like no other. Its mission is unique, and its stunning technological accomplishments, a source of pride and inspiration without equal, represent the best in American skill and courage. At times NASA's efforts have riveted the nation, and it is never far from public view and close scrutiny from many quarters.

However, in all three of these cases, there were indicators of significant issues that were obvious not only in hindsight, and there was strong target fixation from the top down. Over time, this target fixation came at the expense of technical goodness. Specific anomalies and risks were eventually glossed over with little technical rigour – without the rocket science. Our previous successes reinforced greater reliance on 'engineering judgement'. Those judgements sometimes led us to accept some anomalies we didn't understand: if we'd seen them before and they hadn't caused a problem, we may accept them as 'in family' (that is, consistent) with previous performance, even if we couldn't explain them.

These weren't working-level engineering or real-time rocket science errors. The target fixation and the resulting pressures to keep going came from senior managers, and the workforce tried to deliver. The senior managers and executives in each case were experienced and deserved the reputations they had earned as proven veterans in such a tough line of work. They were also responding to pressure from stakeholders to increase the pace and reduce the costs.

Having spent 25 years in Mission Control, many of them at the executive level, I know for a fact that every one of these leaders considered themselves responsible for protecting the astronauts. *Their intentions were good* in every case.

Something clearly matters beyond just hitting our numbers. The discipline we lose in our management reduces the technical rigour in our decision-making and risk management. Veteran, respected leaders were not enough, nor were their professed shared values and good intentions. Top-down management practices still masked concerns and warning signs, and their teams worked hard to meet the leaders' expectations.

And in each of these three cases, the men and women who were trusting us with their lives died.

- NASA's experience in human spaceflight offers tragic examples for how our management practices and the resulting cultural norms not only affect the management team but also influence critical decision-making at all levels.
- There were common themes leading up to each of the three accidents in which astronauts lost their lives under our care.
  - NASA was managing large enterprises, with cumbersome government management practices and a wide range of technical risks, and was achieving success after success in the exceptionally unforgiving environment of space.
  - There was strong, top-down cost and schedule target fixation that came at the expense of technical goodness in managing some risks.
- Over time, specific anomalies and risks were glossed over with less technical rigour.
- Previous successes reinforced greater reliance on 'engineering judgement' and experience at all levels.

▸ The good intentions alone of proven senior managers are not enough to prevent top-down management practices from masking concerns and warning signs that precede failure.

CHAPTER 6

The Leader's Perspective

Every institution is vulnerable, no matter how great. No matter how much you've achieved, no matter how far you've gone, no matter how much power you've garnered, you are vulnerable to decline. There is no law of nature that the most powerful will inevitably remain at the top. Anyone can fall and most eventually do.

Jim Collins, *How The Mighty Fall* [32]

Into the Management Cloud

What the hell?

As we move up into the management ranks, what could cause generations of accomplished, top performers to conduct themselves in ways that are in such conflict with the strongly professed culture and the values still demanded of the workforce? How could the leaders of this organization, who wear The Foundations of Mission Operations on their sleeves, replace the real-time morality with diplomacy over clarity in their decision-making and behaviour?

Obviously, our perspectives change as our jobs change. As the saying goes, 'What you see depends on where you sit'. How so?

The easy and obvious way of thinking about this is where we sit on the organization chart. We promote strong performers from the working level into supervisory positions. Standout supervisors are promoted again into middle-management positions (the white boxes in the MOD organization chart from Chapter 4). Middle managers are then promoted into the senior-leadership positions of Deputy Division Chiefs, Division Chiefs and the executive staff (as shown in the grey boxes).

As we make our way up the organization chart, our responsibilities and authority are typically along the lines that connect us up and down the management chain. Our immediate boss or supervisor is connected directly to us as the next box up on the organization chart. From our boss, we are held accountable for our area of responsibility down the chain to the boxes that connect to ours.

Those same lines of authority also tend to define formal lines of communication. Direction flows down the line from senior managers, through middle managers and to the workforce. Likewise, reporting flows up the line from the working level to each next level of management. Thus, internal to MOD, normal lines of communication are up and down each division's management chain.

The effects of managers' perspective changes become more apparent when we consider how our specific responsibilities change as we move up the organization chart. As the responsibilities change, so do the pressures we feel as we do the job.

For our working-level people who sit in the Mission Control Center, the perspective is dominated by the rocket science and the related real-time risks that ultimately fall only to them to manage. As we saw in Part 1, in real-time we must know the physics, our formal area of responsibility, how it overlaps with other members of the team's areas, and a variety of space sciences. We have the astronauts' well-being, wants, needs and moods to look after, which also add pressure as we make decisions that affect them. Similarly, we feel pressure from Flight Directors who lead the team. They tend to be hard-driving, type-A personalities, who are expected to hold the bar high in individual performance and deliver perfect team performance in support of our core purpose.

Not much else matters in real-time. Politics, long-term financial performance, personnel management etc are all things that don't burden the team as they're managing our most critical risks. And we don't want the real-time team cluttering up their thinking with those things anyway. We need them focused on the real-time morality and keeping their decision-making rigidly aligned to it. Thus, the rocket science and the real-time morality dominate their reality.

As our work moves out of real-time, it also moves out of the control room. Outside of real-time, because we're not managing risks that can literally blow up right now, we have more latitude. We can spend more time in reviews, consider more alternatives, and explore ideas that may not yet be ready for the real world. In these deliberations, we may intentionally consider many of the same things that we intentionally ignore in real-time decision-making (politics, financials), taking into account our customers' less technical needs.

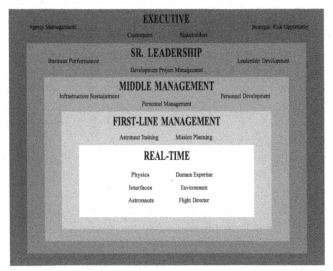

The management cloud

Moving further from real-time as we move up into the first two levels of management, we find ourselves preoccupied with the necessities of managing people and all of the things (offices, equipment, facilities etc) we use to do the job. Although they still review technical work produced by the workforce, first-line supervisors and middle managers spend a large part of their time overseeing the full range of human-resource responsibilities that come with employees. They also manage the wide range of sustainment efforts required for our infrastructure, such as maintenance, repairs and upgrades.

As we move into senior leadership, we leave more of the 'core' work behind. The focus shifts towards the financial performance of the division, development projects (for new or upgraded infrastructure) and leadership development of the middle managers whom we are grooming to replace us in the senior ranks.

The executive level in MOD is comprised of the Director and Deputy Director of Mission Operations and their immediate staff. At the top of the organization, we find our time taken up by top-level business performance, weighing risks and opportunities for various strategic decisions (largely associated with infrastructure and customers) and managing relationships with senior executives from our management chain, our customers and stakeholders. For MOD, this is comprised of the Johnson Space Center Director, NASA Headquarters, MOD's customer-programs (for example, Apollo, Shuttle and ISS) and influential stakeholders in other parts of the federal government.

Just like that, we have moved up the organization chart through middle-level management, senior leadership and the executive level. We have also moved several levels up and into the management cloud, where our perspective and daily decision-making shifts significantly away from the issues that dominated our experience at the working level.

This evolution through the management cloud to the executive level is not unique to MOD. Look again at the management cloud diagram on page 139. Replace the rocket science-related tasks at the working level and in first-line management with any company's product focus, and the responsibilities look awfully familiar. As we move upward through the management cloud, our existence gradually shifts away from working-level production, core purpose and immediate risk management. Instead, it becomes more about managing the enterprise (for MOD, a shift from rocket science to managing $650 million per year and 3,000–5,500 employees). Rather than tactical moves to prevent immediate catastrophe, our responsibilities shift more and more to longer-term strategy, preparing the next level down to move up and replace us in running the organization and managing the business.

In fact, as an enterprise grows larger and managers move higher into the management cloud, the less we are able to directly help our people closer to the real-time rocket science. Those closer to the risks are typically better able to stay up to date with technical details and changing environments. Further, many risks require immediate decision-making and action, and the real-time team must be ready to handle every time-critical risk correctly, every time.

That means, when it comes to directly helping our workforce manage real-time risks, those of us up in the cloud are left to ensure the working level is well prepared, especially for the time-critical situations. We can also review lessons learned, catch process errors and improve processes to eliminate potential escapes and accidents. But our real-time people had better be ready, because we can't be there to fly it for them while we're also managing the enterprise. Likewise, no matter how good the workforce is in the rocket science, they can't manage the enterprise-level, strategic risks from the Mission Control Center.

Nevertheless, no matter how high we move into the management cloud, how lofty our perspective, or how much of our time and attention is dominated by the demands of the cloud, we are still each responsible for *all* of the working-level decision-making and actions within our authority. In any organization, the senior executive is fully responsible for the real-time. Further, that executive position and the entire management team exists to ensure perfect real-time execution. In MOD's case, the Director of Mission Operations, as the senior executive, in addition to managing the $650 million enterprise, is responsible for every action taken during every normal and emergency situation in flight.

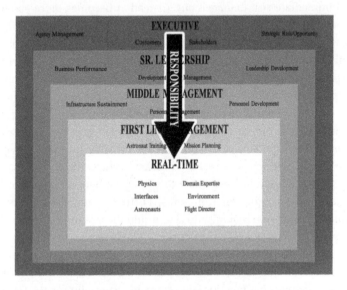

Responsibility in the management cloud

Our perspectives change by necessity as we move higher in the management cloud and shift towards business, strategic risks and personnel management. Rather than tactical moves to prevent immediate catastrophe, we focus more and more on business and management functions.

Our lines of responsibility, authority and communication, like the perspective changes and responsibility shifts, aren't failings. They are reality in management. They also set the conditions that can lead us astray.

▸ Our perspectives and lines of responsibility, authority and communication change by necessity as we move higher in the management cloud and shift away from our core work towards business, strategic risks and personnel management.

▸ Although these changes are reality, they also set the conditions that can lead us astray as managers.

▸ Nevertheless, no matter how high we move into the management cloud, how lofty our perspective, or how much of our time and attention are dominated by the demands of the cloud, we are still each responsible for all of the working-level decision-making and actions within our authority.

A Clouded View and Shades of Grey

With the longer lines of authority and communication, and the perspective changes in the management cloud, comes a change in focus that can obscure the leaders' views despite our better intentions. Therein lies the lurking problem for managers, and there are several contributors to this blurring of our vision.

1. The Lines of the Organization Chart Can Limit and Skew Communication

As we saw with the perspective changes, the organization structure itself is an obvious contributor. As information passes from level to level, it is natural for us to summarize and filter it for concise communication. However, this filtering can dilute the message after passing through layers of management, and can sometimes introduce error. This works both up and down the chain. Managers struggle to keep an accurate understanding of the actions and risks at the lower levels, and the workforce struggles to understand management's goals and directions. This unintentional skewing of the picture is simply the telephone-game played out in the management ranks, but it isn't the only effect of the lines of communication.

In the same way we normalize communication up and down the lines of authority, we can inhibit communication across or outside of those lines. This may start with a natural goal to ensure effective and accurate reporting up and down the chain as an obvious tool for managing risks down our lines of responsibility. Over time, particularly when managers are under pressure, the emphasis can evolve unintentionally towards *concerns* with reporting (*leaking*) risks outside the lines. Managers' concerns with this kind of 'leak' may be well intended at first, perhaps to give them time to understand and respond to a problem. However, the habit can grow into a bureaucratic desire to control, spin or even mask some information to avoid looking bad, protect our turf etc. In this way, it can become normal not just to communicate up and down our own line but also to *prohibit* communication outside those lines.

Consider the MOD management practices in 2004 summarized at the end of Chapter 4. In an organization in which parallel divisions are seen as competing business units and the top-down value is 'no ripples in the pond', any information getting out that isn't controlled from the top is seen as a risk to the senior leadership. Further, if our interest as leaders shifts to simply 'hitting our numbers' and revealing as little of our risks up the chain as possible, even communicating up and down the chain is seen as a risk at each level of the organization chart.

With just these effects, the greying of the management cloud begins. Communication between levels of management loses accuracy, and other pressures cause us to dilute (if not withhold) information. Messages, direction, priorities and goals are misunderstood or not communicated at all.

....

The Emperor Has No Ripples

In early 1995, as an MOD working troop, I was summoned to a lunchtime meeting with the Deputy Director of MOD and working-level representatives from across the organization. This was the continuation of a practice Gene Kranz had followed as Director until he retired the previous year. The instructions were straightforward: 'You are invited to share your concerns and ask the Deputy Director about anything you like. *Nothing is off limits, and there will be no attribution.*' [emphasis added]

I had no idea how I'd been selected, but I was happy to have the attention of the Director for a few minutes. For the previous two years, I had chaired an influential technical panel that was preparing MOD for International Space Station construction and operations, and I was not happy with how some things had changed when NASA altered the nature of the space station program into a partnership with the Russians. The problem on my mind wasn't the Russians but the new NASA program office's implementation of integrated product teams for matrixed management.

Most of my peers showed up with questions about transferring to other areas of MOD and the variety of food from the vending machines. I spoke last and went down a list of examples criticizing MOD's relationship with this new program- management team. In particular, only working-level participation was allowed, and only in the working-level forums. Not only were MOD's senior engineers and managers not represented in higher-level management forums, but they were not even permitted inside the conference rooms to participate from the gallery.

My 'get off the stage' comment was that MOD was not serving this new human space program in the way we had served every one before. We were sending junior, working-level people, most of whom had zero experience in real-time operations, to give engineering recommendations with no vetting from experienced spacecraft operators or MOD management. 'We need the Director to engage with the Program Manager to ensure MOD is bringing the full breadth and depth of the organization's experience base to bear for this important customer, not just the opinions of a small group of well-meaning but inexperienced working troops. As it is, MOD is not supporting them as a directorate. We're just sending them people who happen to have our mail code.'

I could tell I had got the Deputy Director's attention when he said, 'I want the Director to hear this. I'll schedule time with him and let you know when to be in his office.'

Less than an hour later, my Division Chief showed up in my office and asked what in the world I had said to the Deputy Director. After telling him, he said, 'Why would you tell him that? Whether it's right or not, what do you hope to accomplish by telling him that? I wouldn't tell the Director that.'

Minutes later, the MOD executive responsible for all Shuttle and International Space Station operations (and my Division Chief's boss) came by to have the same conversation with me. His final comment was, 'Well, you're right about our engagement with the new program office. But what do you hope to accomplish by telling the Director that? You can do what you want, I won't tell you not to, but I wouldn't tell him that.'

Believing it was my job to speak up, I did tell the Director the next day in his office. He sure seemed to get the message and share my concerns.

Afterwards, a veteran Flight Director and mentor called me into his office and asked me to tell him what I had said to the Director, now that my name was making the rounds among MOD's senior leaders. After hearing my story, he said, 'Look, you need to stop saying things like this. It doesn't matter if you're right. The way it works in MOD is that you have to earn the right to say some things and criticize how we do business. You haven't earned that right yet. So, if you keep talking like this, MOD management will end your career for you. Look, do you want to survive long enough to fix this stuff? Then don't forget any of your concerns – just shut up about them for now. When they promote you high enough, dust off your list and fix things. Until then, shut up and lay low.'

So much for 'no attribution'! I begrudgingly took his advice.

What error in my message was my management chain concerned about? None.

In what other ways were these senior leaders working to resolve the problems *they agreed* existed in our relationship with the new program and customer? None were explained to me.

The focus was simply on not stirring things up – not putting ripples in the pond. Rather than risking that, let's just keep this to ourselves and not inform the boss. That priority seemed to override whatever concern they shared that MOD was not effectively serving the next large human space program, which happened to provide half of our budget. It seemed to be just as significant to them that I was seen as way out of line, figuratively and literally, and my senior managers did not want the topic even mentioned. It didn't matter if my point was right. As my mentor said, I hadn't 'earned the right to say it'.

2. Alignment to Values is Diluted by the Less Catastrophic Ramifications

As with the perspective changes, the lines of authority explain only part of the 'greying'. For the rest, we return to the other aspects of the management cloud. As we do, consider the challenges from Chapter 2 that make some things difficult for the workforce. These are still with us even as managers and leaders:

- Technical complexity.
- Incomplete data before a decision is required.
- The stakes and ramifications of being wrong.
- The 'human element'.

The first two challenges are not really much different for us as managers in any organization. Sometimes we're faced with complicated subject matter and wish we had more time to make a decision. However, the stakes and ramifications of being wrong in the management cloud are typically not as obvious and potentially catastrophic as they are in Mission Control in real-time.

We saw in Part 1 that the real-time morality and the potential for immediate and obvious catastrophic failure are tremendous motivators to keep us aligned to core purpose, willing to overcome our discomforts and fears, and to say what needs to be said while flying the rocket and spacecraft. But as we move up in the management cloud, we find our daily work and decision-making further and further away from the technical and real-time world of the Mission Control Center. Our attention is filled with schedules, personnel, budgets, strategic decisions, stakeholder relationships . . . Rarely does an executive decision result in blowing up the rocket we're flying *right now* in Mission Control.

Thus it is inherent in management roles that the aligning and empowering influence from the worst ramifications of our mistakes is diluted. It then becomes easier to rationalize decisions without demanding the same alignment and the same obsession to understand why each is the right action. It becomes easier to unlearn the behaviours that were so important to preserve the Mission Control trust and to deliver highly reliable decision-making. Because, 'We're not going to blow anything up with this decision.' This is management, not rocket science, right?

Or, as an MOD executive told me when I was a Flight Director leading Mission Control, 'I wish my job was as easy as yours. Management at this level is nuanced, and you just can't understand it until you've experienced it.'

3. Responsibilities Shift to the Discomforts We Are Frequently Not Prepared For

That diluting effect in aligning to core purpose leads us to the last challenge from Chapter 2, the human element, which also evolves and contributes to the greying effects. As managers and executives, we still have those discomforts to contend with where we ask ourselves things like, 'Who am I to make this call? What if I'm wrong, and I'm to blame?'

Look back again at the management cloud diagram. Our day-to-day decision-making becomes less about the rocket science and real-time risk management that got us promoted. It has to, since our responsibilities evolve towards the less empirical or rule-based management functions that are also frequently foreign to us. Those changes can add to the working-level discomforts, for instance, 'This ain't what got me here. What if I'm not good at it? Can I still make a difference? How will I get promoted again? Am I being set up to fail?'

Once in this foreign territory, we find ourselves in positions that require us to pass judgement as managers. We judge our employees' performance, the progress of various projects, the merits of new investments, and more. More challenging still, we may find ourselves leading discussions that result in changes to 'the way we've always done it and know to work'. Great. Now, what if people don't like my judgements? Maybe they won't like *me*, which certainly won't help me do a good job or get promoted.

Our responsibilities in the management cloud can also carry a larger share of the human-element challenges associated with influencing people and managing relationships with customers and stakeholders. Overall, rather than the more black-and-white subject matter faced in real-time at the working level, in the management cloud we find ourselves with greyer situations and less obvious right answers.

The combined effect of the new responsibilities, the need to pass judgements and lead change, and the vagaries of influencing and managing relationships (compared to rocket science) all add to the human-element discomforts for many of us. Just as they affect the working level, the discomforts increase the intimidation we feel in the face of the stakes and ramifications from every decision. Without specific development to help learn skills and behaviours to be effective in these new roles, this increased intimidation can get in our way.

4. We Often Hide Behind Previous Success to Avoid New Behaviours and Discomforts

Many of us revert back to personal styles that brought us success before at lower levels of the organization. Those behaviours made us successful then, so we expect that they will again as we find ourselves with human-element challenges we're not comfortable with. As a young Flight Director, my boss reminded me to focus on the strengths that got me my job, and the need to 'dance with the one who brung you'.

So we find ourselves in new roles, frequently reluctant to give up our previous roles and behaviours, and either reluctant or ill equipped to step up to the new one. Since our superiors came up this same way, we may also find ourselves surrounded by senior leaders who not only share the same discomforts and intimidations, but who will also actively dismiss all but the working-level work – the rocket science. Over time, more and more of the leadership team is comprised of people who stood out at the working level, are less comfortable and experienced at the more 'touchy-feely' roles that dominate their management responsibility, and are not led through a process to grow into those roles in the same way they were when they were indoctrinated into the real-time morality.

A Flight Director colleague summed up this typical MOD management attitude as he was transitioning into his first executive position: 'The most difficult things you'll ever need to learn are the things you learned to become a good Flight Director and lead missions. Everything else you need to know to be a manager you can learn in the first couple of weeks on the job.' (A year later, the barely experienced executive was just as emphatic about how much he did not know about managing people and budgets, and how much more there was to managing an enterprise than just getting the rocket science right.)

5. We Trivialize Inconsistencies as the Nuances of Management

Summing it up, thanks to the perspective changes and the effects of the management cloud, we move into increasingly non-technical and human-element related work that we are less prepared for and comfortable with. At the same time, we feel less and less direct connection to the immediate ramifications of failure, diluting the 'organic' alignment from the real-time morality in our management decision-making. As a result, rather than taking action as deliberately in our management roles as we do at the working level, the higher we go into the management cloud the more likely we are to dismiss disconnects and inconsistencies in our decisions as the 'nuances' of management.

This leads to exactly what Allen Flynt found in MOD in 2004: leaders who were comfortable with rocket science but much less comfortable and developed in their management roles, and unable to discuss it because of the cultural norm to keep ripples out of the pond and focus only on the technical challenges. And they were raising their successors in that same image, because to do otherwise would require talking about the things that we don't really value and aren't comfortable talking about anyway!

These effects all grow, not only as we progress up through each level, but also as the enterprise itself grows. For example, in comparison to MOD leaders' challenges, the executives managing our multi-billion dollar customer programs (e.g. Apollo, Shuttle and ISS) must deal with longer lines of authority and communication and shift their perspective for enterprises that are 5 to 10 times the size of MOD.

▸ There are several contributors to this blurring of our vision in the management cloud:

    1. The lines of the organization chart can limit and skew communication.

    2. Alignment to values is diluted by the less catastrophic ramifications.

    3. Responsibilities shift to the discomforts we are frequently not prepared for.

    4. We often hide behind previous success to avoid new behaviours and discomforts.

    5. We trivialize inconsistencies as the nuances of management.

▸ These effects all grow, not only as we progress up through each level, but also as the enterprise itself grows.

Errors and Risks

The clouded view and the shades of grey in the management cloud dilute our alignment to core purpose and the values of the real-time morality. That leads to perceptions and behaviours that can add risk and errors to our decision-making.

As new managers, we may see inconsistencies between our thought process, value judgements etc compared to our previous experience at the working level. While at the working level our tendency is to pursue and resolve those inconsistencies, as managers we come to see these inconsistencies as nuances and accept them as part of our new reality. That gives way to not seeing the inconsistencies and then defending them as normal in the management ranks. If we don't do something about that evolution to normality, our leadership can go down a path like the one that led to MOD's 2004 management practices, as described in the tables in Chapter 4, and then outcomes like we saw in Chapter 5.

Like the perspective changes in the management cloud, this evolution is gradual. It starts with normalizing the weirdness.

Normalizing the Weirdness

When we start a new job, it is natural for us to focus first on learning the new rules and fitting in to our new role. This is also true for us in new management roles. We still want to fit in and be part of the team. We want to be effective engaging with peers and bosses. We want to be seen as part of the solution.

An element of this transition is learning and adopting the expected behaviours in order to be accepted and be effective. Some of these new behaviours may appear odd to us at first. In some cases, because of the effects of the management cloud, we may find management practices that appear inconsistent with the behaviours we learned so deliberately at the working level. As described in Chapter 2, these include working level behaviours such as open and direct communication, alignment first to our core purpose, technical truth, integrity and courage. Or, as we saw in Chapter 4, we may find a switch from 'clarity over diplomacy' to 'diplomacy over clarity' in the management cloud.

Over time, as we become more comfortable with the management practices and familiar with the inconsistencies, we find ourselves accepting the weirdness – or what many of us are able to perceive as weirdness when we first move out of the real-time environment into management. This may start as a desire to fit in. It can easily turn into a matter of survival as we also try to impress the boss and remain promotable in the eyes of the senior leaders, which just reinforces this normalization.

Along the way, many of us lose the ability to see the inconsistencies, the weirdness. Not only does this happen with little or no challenge from many of us, but it also becomes 'normal' as we accept that life in management is 'nuanced'.

....

What Can a Poor Manager Do?

During a brief stint as a first-line supervisor in 1995, one of my top performers came to me with her concerns about the state of MOD's individual-award process. In order to break an appearance of favouritism, we had moved towards peer-nominated awards. Thus, working-level people nominated other working-level people for recognition. A committee of other working-level employees from across MOD then evaluated the nominations and selected awardees, and there was no management participation.

My employee (who was on the committee) was concerned that some of the award recipients were widely seen as undeserving and that we were at risk of encouraging apathy from the devalued awards. Her main point was, 'The only objective criteria for a work performance award are quality and quantity of work. No other criteria matter.'

This excerpt from a committee email was the prompt for her note to me:

> We need to encourage more of the *deadwood* [emphasis added] and average-Joe nominations in order for this to become a peer award. The superstars are going to get their awards (rewards). How do we encourage awards for the average Joe who does an occasional good thing?

I shared her concern and forwarded her email to our Division Chief. This was the same senior leader who only a few months before had told me, 'Whether it's right or not, what do you hope to accomplish by telling him [the Director] that?'

This email elicited a similar response from him. He showed up in my office and asked what my point was. I simply repeated the points of the original email:

> Isn't this a basic management responsibility . . . setting performance standards, awarding top performers? How would working-level employees be in a position to judge performance across organizational lines in areas in which they have no experience? Do we really want to send this kind of message – that we should be awarding more of the 'deadwood' as an incentive, rather than singling out model performance? Again, is this what the management team really wants? Where do we think this will drive our culture?

The Division Chief responded, 'Look, Paul, culture is what it is. There isn't anything we can do about it as managers. All we can do is try to understand it and not impact morale.'

I hadn't yet assimilated, I guess. I added it to my list of things to shut up about until MOD promoted me some day. After this episode, that seemed even more unlikely, again.

We may start with a willingness to shrug off confusing or unpopular management decisions as unimportant, because they're not going to blow anything up, or we're just giving in to our discomfort. However, from there, it's an easy step to not seeing the weirdness (or the cognitive dissonance, in psychological terms) and buying in to the 'nuance' of it all.

Then we progress to defending it, nuance and all. This culture in the management ranks becomes the thing we must now master in order to fit in, perform well, and move up. As we adapt our behaviours and fit in to the expected management practices, we develop a sense of ownership that comes with authority, 'Hey, I was promoted for a reason, you know. You don't understand how it works because you're not at this level . . . where it's nuanced. This is the way we have to do it.'

....

Making the Slope More Slippery

In his career as one of the first Flight Directors in history and then Director of MOD for 11 years, Gene Kranz was a dominant figure in establishing and stewarding the Mission Control and MOD culture. He was not shy about reminding the management and the workforce about our responsibilities; essentially that the real-time morality trumped everything.

As an indicator of how quickly things can change, consider this comment to the Flight Directors in 1997, just three years after Gene retired. One of the executives Gene had groomed for the job told us shortly after he became Director:

> When I gave my first speech to MOD as Director, John O'Neill saw me walking on stage with note cards, like Gene used to do. I was prepared to do my best Kranz imitation. John said, 'Put those away. You're not Gene, and you don't need to talk that way. Just be yourself.'

And you know, he was right. Gene was good back in the day, but things have changed. Now, we have to be willing to go along and get along. Gene wouldn't survive today. Besides, we don't need that kind of leadership any more. It's a different world.

....

This wasn't just any group the Director was saying this to. This was the entire group of Flight Directors who were leading Shuttle flight operations and preparing Mission Control to construct and operate the space station in orbit. We were hearing these words from a Director who had been the Chief Flight Director.

This wasn't about Gene, but about deliberate leadership in a tough business. 'Go along and get along'? Where the hell is that in The Stone Tablets? How does that fit with the real-time morality? Go tell that to the astronauts on the next mission!

Incidentally, this new Director was the same executive who, like my Division Chief, had told me, 'Well, you're right . . . But what do you hope to accomplish by telling the Director that? . . . I wouldn't tell him that.' Since then, he had progressed higher into the management cloud and now saw his challenges through a more nuanced lens than the real-time morality. Rather than tackling his own difficulty in stepping up to the various discomforts in the management cloud, he had accepted that the values just didn't apply in our different world.

As the most senior executive in MOD, he influenced more than just the Flight Directors. He set the tone for all of our management discussions. Over the next several years, he would drive MOD's management practices towards no ripples in the pond, a phrase he believed reflected the sensitivities of our different world and would keep MOD out of trouble. To his great credit, he later realized this mistake and was instrumental in bringing Allen Flynt into MOD to undo it.

Serious Ramifications

Of course, the risks we worry about as managers aren't principally about leadership style and management practices. They are about mistakes and failure when the stakes are high, and the ramifications are severe. However, it is the loss of alignment to core purpose and values, and the 'normalization of the weirdness' in the management cloud, that prevent us from seeing preventable mistakes that lead to failure.

Although our management decisions don't typically have the same risk of immediately blowing up the rocket, there are still severe ramifications that are directly related to our conduct in the management cloud. Most obviously, we can make poor decisions that cost the organization money. Whether that money is lost because we fail in our mission, or because some investment, development project or mission exceeded the cost estimates, it still represents lost opportunity to accomplish something else. That money has to come from somewhere, and if we string together many of these losses, we can severely impact the organization's balance sheet. In industry terms, we cost the company profits. We could cost a customer money, which could also cost us the customer, worsening the impact on the company. For MOD, if we exceeded our budget for the year, it would come at the expense of some technical goal our customer programs otherwise sought to achieve but would then be forced to scale back to pay for our lack of cost control.

Further, we can put people we are responsible for at risk – if our poor decisions cost the company too much, we could bankrupt the company and put our enterprise out of business. Similarly, in MOD's case, we could contribute to cost overruns that cause NASA to cancel a program. In either case, our people's livelihoods are at stake in many of our management decisions.

In other cases we manage similar risks not from the financial impact but simply from the damage to the organization's reputation. It does not take many days of critical headlines when NASA is chasing a rocket engine problem or bad weather at the launch pad before leaders start feeling pressure to solve the problem and get on with things. We don't want to let down the leaders who came before us and were able to do hard things. We also don't want to lose the confidence of our stakeholders.

That too appears trivial at first blush, but this was exactly MOD's situation in 2010, when NASA cancelled the Moon and Mars program Constellation, as described in Chapter 5. Although the cancellation was attributed to cost and politics, ultimately NASA Headquarters lost confidence in the field centers' ability and willingness to innovate. The result was laying off tens of thousands of experienced technical workers from a space agency that retained the same overall budget! NASA chose to route much of the funds elsewhere.

In business terms, an extreme example of the ramifications from a damaged reputation is found in the accounting firm Arthur Andersen. In 2001, they were a $9 billion company with 85,000 employees worldwide. In 2002 the actions of a small number of employees in the Enron scandal resulted in felony convictions, and Arthur Andersen surrendered its CPA licence. Although the company's conviction was overturned by the US Supreme Court in 2005, the damage to their reputation as trusted auditors was irreparable. They were down to only 200 employees and were not able to return to business.

Impacts like business losses, bankruptcy and going out of business still may not compare to blowing up the rocket to some of us. Consequently, the management cloud tempts us to shrug them off as management risks that aren't catastrophic. That temptation ignores the ultimate consequences that executive mistakes can lead to, and have.

That is how MOD gradually evolved into the management practices described in Chapter 4. A management team comprised of top rocket science performers lost perspective and connection to our most fundamental purpose and values in practice. In spite of our professed focus on always getting our rocket science right, 'no ripples in the pond' and 'keeping cards close to the vest' eroded not just our management practices but also found their way closer to our rocket science.

At a higher level in NASA's human space programs, that is also how target fixation led repeatedly to accepting risks with little or no technical rigour and then directly to catastrophic failures and the deaths of astronauts. Like MOD, veteran, respected leaders who professed shared values and good intentions were not enough to prevent top-down management practices from masking concerns and warning signs that preceded failure.

As already noted, as we move into higher management our responsibilities and attention are consumed with the real and necessary management commitments of running the enterprise. In parallel, we widen the separation between our daily decision-making and the more immediately catastrophic ramifications of our decisions. In turn, we risk gradually losing alignment to the core purpose and values that are key to highly reliable performance and success in managing the catastrophic risks. We can then remain either unaware or unwilling to address the negative effects of our management practices as they lead us to preventable failure in the core product the enterprise exists for in the first place!

In MOD terms, that could lead to sending unintentional messages from management to the workforce that dilute the focus on protecting astronauts, truth, integrity and courage – that some other management value takes precedence over the real-time morality. At a higher level, we may challenge our rocket scientists to reconsider their concerns with catastrophic failure in light of our schedule or budget milestones.

▶ This is the dangerous nature of the management cloud and its influence on leaders:
  ▷ As we move into higher management, our responsibilities and attention are consumed with the real and necessary management commitments of running the enterprise.
  ▷ In parallel, we widen the separation between our daily decision-

making and the more immediately catastrophic ramifications of our decisions (we're not going to blow up any rockets).

▷ In turn, we risk gradually losing alignment to the core purpose and values that are key to highly reliable performance and success in managing the catastrophic risks.

▷ Many of us lose the ability to see the inconsistencies, and then buy in to the 'nuance' of management – we normalize the weirdness.

▷ We can remain either unaware or unwilling to address the negative effects of our management practices as they lead us to preventable mistakes and failure in the core product the enterprise exists for in the first place!

It Is Still a Morality

The ultimate moral issue isn't about indefensible management practices and culture, but *their effects on managing the fundamental risks* for which the enterprise is responsible. For MOD, 'no ripples in the pond' found its way closer to our rocket science, threatened our ability to preserve the real-time morality, and cost us the confidence of our customers. At the higher levels in NASA, the repeated cycles of accepting risks with little or no technical rigour led directly to catastrophic failures and the deaths of astronauts.

Applying the same judgement we apply in the Mission Control Room, MOD's leadership failure was succumbing to the diluting effects of the management cloud, normalizing the weirdness in our practices, and *not* deliberately leading our organizations in practices that were aligned to our core purpose and values. Consider the norms that became more important than relentlessly upholding the integrity, technical truth, courage and alignment to core purpose in our decision-making. Which management norm trumps doing the right thing? No ripples in the pond? Schedule pressure? Respect for a veteran leader's judgement and intuition, as in faster-better-cheaper?

Yes, even as managers, the morality is with us. How we answer those questions, and more importantly, how we *manage* and affect the cultural norms based on those answers, determines which side of the morality we are on as managers.

▸ As we move up, managers risk gradually losing alignment to the core purpose and values – the morality – that are key to highly reliable performance and success.
▸ MOD's leadership failure was succumbing to the diluting effects of the management cloud, normalizing the weirdness in our practices and not deliberately leading our organizations in practices that were aligned to our core purpose and values.

Insidious Errors

As managers adjust their personal perspective and normalize the incremental inconsistencies between management practices and the real-time morality, we are also contributing to a broader normalization, particularly down our lines of authority and responsibility.

This happens because, as we make decisions, managers are setting policy and behavioural norms. Sometimes those policies and norms are deliberate. At other times they're the result of signals we may be sending unintentionally as our employees overinterpret decisions and personal style as the boss's preferences and values. Over time, those policies and behavioural norms can become more ingrained and cultural, whether intentional or unintentional. For example: 'no ripples in the pond' in MOD, 'prove it isn't safe' in the Shuttle Program, and the effects each had on management practices and risk management.

Our risk, as leaders, is that someone who is aware of some mistake in our decision-making won't muster the courage to overcome that discomfort and intimidation. As a result, some action will not be taken and someone who could have prevented a costly mistake or offered a new opportunity will not speak up. That kind of intimidation doesn't just present itself in the midst of an emergency or catastrophe. It is there in varying degrees every time an employee has to choose to say something that's different from everyone else.

The most troubling errors aren't the ones that are obvious to some people in the community who must decide whether or not to speak up. In those cases, at least someone recognizes the error, and we have the chance of someone stepping up. They may have to overcome resistance from the community and their own discomfort, but just having some awareness inside the organization of an error leaves us with an opportunity to fix it.

The truly worrisome risks are the errors right in front of everyone that we have all accepted as fact. In a quote apocryphally attributed to Mark Twain, 'It ain't what you don't know that gets you into trouble. It's what you know for sure that just ain't so.' Bias, subjectivity and error creep in, often unintentionally, from a variety of sources and then cloud our understanding and judgement. Previous successes – ours, our heroes', accomplished leaders' – can lure us into thinking we are right, whether we can explain it or not.

At other times we know what *should* happen in some situation. When presented with data that conflicts with those expectations, we sometimes discard the conflicting information rather than consider it an indicator of something wrong or the need for a change in our system. Over time, this same conflicting information may be considered 'proven' to be unimportant if all other indications meet our expectations, and we continue to be successful. Thus, just as leaders 'normalize the weirdness' in management practices, we can normalize the *community's* willingness to discard information that conflicts with our preconceived notions.

As a result, the organization may be completely aware of (and unconcerned by) the conflicting information, whether we can explain it or not. Once accepted, these biases and errors become much more difficult to overcome. They can take on a well-intended and sincere moral imperative in the organization as we defend our biases by saying things like:

- We've always done it this way.
- We don't understand it, but it's 'in family' (consistent with previous experience).
- We all know this is safe.
- Who are you to disagree with this respected leader or that expert?

All of the worst ramifications – and *failure* – could then come to pass from some longstanding bias that has gone unseen, forgotten or unchallenged as 'conventional wisdom'. At all levels of an organization, biases can be so strong that we may not even consider any alternative perspectives, and we therefore won't have the opportunity to catch the error and deliberately manage the risk. Concluding that we were wrong in our initial conclusions feels harsh, like a negative judgement about ourselves.

However, the real-time morality judges us on *how* we decide – based on what information. We are only outside of our morality if we refuse to see our error once we have the additional data. We are not being disloyal by concluding we were initially wrong. We are maintaining the integrity in our thinking and showing the courage to see our error and embrace the more complete technical truth.

▸ Just like in real-time, the morality applies in all decisions, not just in the case of catastrophic failure.

▸ We have this ongoing risk that some action will not be taken and that someone who could have prevented a costly mistake or offered a new opportunity will not speak up – won't muster the courage to overcome that discomfort and intimidation.

▸ The truly worrisome risks are the errors right in front of everyone that we have all accepted as fact.

▸ We struggle to accept that we were wrong in some previous decision because it feels harsh, like a negative judgement about ourselves – to overcome this, we must reframe the judgement:

- We are only outside of our morality if we refuse to see our error once we have the additional data.
- *We are not being disloyal by concluding we were initially wrong* – we are maintaining the integrity in our thinking and showing the courage to see our error and embrace the more complete technical truth – *we are aligned and loyal to our core purpose*.

We Sometimes Don't Manage the Way We Fly

Simply by moving up the management chain, how does it become irrelevant to ask: How do we know this? Why is this the right decision? How does it support our core purpose? How do we explain any inconsistencies with our values?

The management cloud is waiting for us with all of its subtle pressures that lead us to normalize the weirdness. As leaders move up, we must, therefore, be *increasingly deliberate* in compensating for the diminished effects of those readily apparent, immediate ramifications of errors and the increasing pressures in areas that bring discomfort and could inhibit effective action.

Our mistake in MOD's management ranks was in over-applying the observation that management work is nuanced and the values are different from those at the working level. We assumed that since managing isn't flying, the real-time morality did not apply.

As we move up, we must all adapt to the changed responsibilities in the management cloud. The daily jobs and how we do them *are* different from 'just' handling the rocket science. The more human element-related roles we face as managers *are* nuanced compared to the more rules-based work at the working level in Mission Control.

While it is true that managing and flying are not the same, it is the *job* that is nuanced, *not* the morality. We lost that distinction in the management cloud. The management cloud's diluting effect on the real-time morality led us to dysfunctional management practices.

To apply the real-time morality in management practices, leaders at every level would do well to remember the Mission Control mantra, 'Train the way you fly. Fly the way you train.'

▸ MOD lost the distinction in the management cloud that it is the job that is nuanced, not the morality.

▸ The management cloud is waiting for us with all of its subtle pressures that lead us to normalize the weirdness.

▸ As leaders move up, we must be increasingly deliberate in compensating for the diminished effects of those readily apparent, immediate ramifications of errors and the increasing pressures in areas that bring discomfort and could inhibit effective action.

Part 2
Losing Faith
Management Clouds the Morality

KEY POINTS

▶ As we move up, managers' realities are consumed with real and necessary business and personnel-management commitments, and we risk gradually losing alignment to the core purpose and values that are key to highly reliable performance and success.

▷ Many of us lose the ability to see the inconsistencies in our management practices and then buy in to the 'nuance' of management – we normalize the weirdness.

▷ We can then remain either unaware or unwilling to address the negative effects of our management practices as they lead us to preventable mistakes and failure.

▷ These effects all grow, not only as we progress up through each level of management but also as the enterprise itself grows.

▶ We can lose this distinction in the management cloud: it is the *job* that is nuanced, *not the morality.*

▷ Focusing only on 'hitting our numbers' can mask management practices that erode effectiveness over time.

▷ The erosion can lead to dysfunctional management practices that impact our ability to deliver on our core purpose and prevent failure.

▷ Just like in real-time, the morality applies in all decisions, not just in the case of immediate catastrophic failure.

» Over time, managers' policies and behavioural norms can become more ingrained and cultural, whether intentional or unintentional.

» Challenging policy and cultural norms takes courage.

» We have an ongoing risk that some action will not be taken and that someone who could have prevented a costly mistake or offered a new opportunity will not muster the courage to speak up.

- ▶ The truly worrisome risks are the insidious errors right in front of everyone that we have all accepted as fact.
  - ▷ Once accepted by experts and leaders, biases and errors become much more difficult to overcome, despite new data or study that otherwise could have led the team towards truth.
  - ▷ All of the worst ramifications – and *failure* – could then come to pass from some longstanding bias that has gone unseen, forgotten or unchallenged as 'conventional wisdom'.
  - ▷ At all levels of an organization, biases can be so strong that we may not even consider an alternative perspective, and we therefore won't have the opportunity to catch the error and deliberately manage the risk.
- ▶ Our struggle is accepting when we were wrong in our initial conclusions, because it feels harsh, like a negative judgement – to overcome this, we must reframe the judgement:
  - ▷ We are only outside of the real-time morality if we refuse to see our error once we have the additional data.
  - ▷ *We are not being disloyal by concluding we are wrong* – we are maintaining the integrity in our thinking, showing the courage to see our error, embracing the more complete technical truth – *we are aligned and loyal to our core purpose.*
- ▶ Judged by the real-time morality, our leadership failure is succumbing to the diluting effects of the management cloud, normalizing the weirdness in our practices and *not* deliberately leading our organizations in practices that are aligned to our core purpose and values.
  - ▷ The good *intentions* of proven senior managers are not enough

to prevent top-down management *practices* from masking
concerns and warning signs that preceded failure.

▷ How we manage and affect the cultural norms determines which
side of the morality we are on.

▶ The management cloud is waiting for us all with its subtle
pressures that lead us to normalize the weirdness.

▷ As leaders move up, we must be *increasingly deliberate* in
compensating for the diminished effects of those readily
apparent, immediate ramifications of errors and the increasing
pressures in areas that bring discomfort and could inhibit
effective action.

▷ Choosing otherwise is a leadership failure that will erode
management practices and lead to failure.

PART 3

···············

Refinding Faith

Transforming the Leadership Team

It is leadership and demonstrated leadership values that set the tone in your people's willingness to step up and bring you technical truth – before the universe does in a way you cannot afford.

Paul Sean Hill, Director of Mission Operations, 2013

CHAPTER 7

Leadership Evolution
Ideas that Inspired

Management makes a system work. It helps you do what you know how to do. Leadership builds systems or transforms old ones. It takes you into territory that is new and less well known, or even completely unknown to you.

John P. Kotter, Leading Change[33]

Rippling the Pond – Awareness

MOD, the proud organization that had developed generations of steely-eyed missile men and women into high-performing, highly reliable decision-makers, who still cherished our leadership culture and legacy, had lost the way in the management cloud. We had lost the ability to apply the same moral judgement to our management practices that the real-time morality demanded of us in our most critical work. That led to the dysfunctional management practices from Chapter 4 that confronted Allen Flynt when he took over in 2004 as Director and thought, *Really? This is MOD?*

Also, recall from Chapter 5 that, when he was given the job he had been told that, like it or not, MOD was facing a loss of confidence from our stakeholders. This wasn't just a matter of principle; Allen had been charged with saving this critical national asset.

Now what? Where would he start?

Allen quickly gave up simply preaching values and hoping to align the management team in the spirit of just 'doing the right thing'. MOD managers weren't interested in hearing an outsider's version of values. No ripples in the pond, hold your cards 'close to the vest', circle the wagons, diplomacy over clarity . . . However we say it, the effects were too well established, as were our resulting management practices, resisting any pressure to innovate and reduce costs. Besides, *we* were MOD – the conscience of human spaceflight – and we didn't need someone else to lecture us about values.

So, Allen looked for ways to rally his managers around a specific common cause. What better way to do that than engage the entire group in assessing MOD's strengths, weaknesses, opportunities and threats? It wasn't a new idea, but it was still the obvious way to identify top priorities. The team discussions should also have helped break down the walls and the silos that had developed as the divisions came to see each other as competitors instead of team-mates. It didn't work. As mentioned in Chapter 5, retreats with the entire management team in 2004 and 2005 resulted in plans to keep studying, and only confirmed the team's existing prejudices: 'We're MOD. We invented human spaceflight. No one else in the world does it but us. We have nothing to learn from anyone else'. None of which did anything to identify strengths, weaknesses, opportunities and threats or rally the management team around a common cause.

Allen discovered how thoroughly his direct reports had given in to 'no ripples in the pond'. Undoubtedly, MOD's effective indoc-trination of these managers when they were at the working level was now working against us: as managers, they had again learned the expected behaviours – now the expected management practices – and they were sticking to them. Besides, they had also all learned that executives and management fads come and go. Many of these same MOD managers privately admitted they were waiting Allen out: 'He'll be gone in a few years too, and we'll still be doing what we've always done'.

Although that was a typical attitude, the ongoing dialogue did get the attention of a few senior managers. These managers heard things from Allen they weren't accustomed to hearing at that level, and were curious about the change he was trying to bring to the management discussions. By 2006, this group included only five of Allen's 17 direct reports, but they were at least open to more unguarded discussion about MOD's values and challenges, and to ideas to reduce our costs. (It was at this time that Allen brought me back to MOD to be a direct report, MOD's Manager for Shuttle Operations, and one of those five.)

In another attempt to change our management focus, Allen tasked his Chief Engineer with taking teams of the senior managers to visit other industries, both government organizations and private companies. His goal was to benchmark their best practices for lessons we could apply. MOD's senior managers humoured the boss. What better way to confirm MOD's superiority in all ways than visiting one management team after another to see the contrast for ourselves? Of course, since 'no one else flew people in space', there was only so much we could really learn, right? It may end up being a waste of time, but the boss said to go. So we went.

As each team returned from a benchmarking trip, they briefed MOD management on their findings. Some common and surprising conclusions got our attention. First was the recognition that a wide range of industries managed challenges similar to MOD's:

- Large capital investment in core infrastructure.
- High-energy systems that could fail catastrophically.
- Personal risk to the workforce and the public.
- Situations that demanded continuous highly reliable decision-making.

Further, none of the benchmarked organizations had anything to do with human spaceflight, and only one of them worked in space at all. These included the Southwest Airline's operations center, Arizona Public Service power grid control, New York City's emergency control center and the Iridium satellite control center. To our great surprise, MOD was not way out in front of these organizations in all areas, and in some cases we appeared to be way behind widely used standard practices. Worse, we found specific instances in which these outside organizations were more effective in areas that MOD considered ourselves world leaders, such as training management, analytical tools and large data network design and management.

More than any specific practice we saw, gaining this unhappy awareness was the single most valuable lesson MOD learned in our benchmarking. Most of us now shared an undeniable concern that we could be behind in practices we considered close to our core. It's one thing for MOD to not be the greatest financial strategists, but training, data analysis and MCC-like networks? It could only have been worse if we had discovered we did not understand rocket science or real-time operations!

This dawning awareness was the perfect set-up for benchmarking the USAF 50th Space Wing in May 2006, less than a year into this practice of looking outside our walls. These people operated more than 60 military satellites that were critical to national security. They were responsible for mission planning, training and operations for space systems, with more than 1,600 people in several control centers. Many of us found their mission and work environment to be surprisingly similar to our own. Once again MOD found areas in training and in control center technologies in which we were far behind. We also took note of the smaller relative size of the 50th Space Wing's training and operations teams compared to our own.

Unlike the other benchmarking trips, the visit with the 50th Space Wing included most of Allen's direct reports. Thus, we'd all been there to make these observations for ourselves, not just hear them in a summary briefing. However, old biases are tough to overcome. The team had seen many things that applied to our work, and many things we could learn, but there was still a familiar reluctance to buy in, because these satellite jockeys 'still aren't *manned* spaceflight'. After hearing that comment a number of times from the team, I was disappointed, and told them so:

> What are we talking about? When are we going to stop hiding behind ideas like 'no one else does what we do'? Every time we do that, we convince ourselves we have nothing to learn from anyone else.

> How many more do we have to find who look just like us before we stop saying that? Don't you think that maybe they have something to teach us after all? How is it a weakness for us to learn from them or anyone?

> How many times do we have to see evidence that we're falling behind in areas we think we're best in before we accept it and do something about it? How many times are we able to see that we're behind in *anything* without choosing to do something about it?

Our management discussions changed over the next several months. After stewing on our benchmarking findings, most of us gave in reluctantly to the observations. That reluctance stemmed not only from the long-cultivated norm of 'no ripples in the pond' but was also partly driven by a strong sense of loyalty to MOD. Criticizing rather than defending the organization still felt disloyal to many of us.

However, we could see that our management practices didn't compare as well as we'd like to other operations/organizations. No longer could we hide behind our unique history and ignore the contrasts. As the diagram on page 176 shows, once we accepted that even organizations without our unique responsibility shared many of our management challenges, we were able to make many more direct comparisons and see shortcomings in some of our practices. Fortunately, those same comparisons offered us solutions through the other organizations' experience and best practices. We could incorporate those outside ideas into our own situation without giving up who we were or lowering the bar on the performance standards in our core work. Rather than disloyalty, it was our responsibility to grapple with these comparisons and work to improve our own management practices as loyalty to our core purpose and our people. With that, most of MOD's senior managers understood the need for us to make some changes.

This growing awareness by MOD's senior managers was the critical first step to doing something about the observations we didn't like from our benchmarking, and to looking more critically at all of our practices. Our challenge now was to decide what to do, and then do it.

TEAM AWARENESS GAINED FROM BENCHMARKING AND SEEING THE RIPPLES IN THE POND

By looking outside our walls we found relevant management lessons, even from very different industries and organizations

WE'RE NOT AS UNIQUE AS WE THINK WE ARE

This enabled us to see contrasts in our practices that were not best, world class, or even up-to-standard practices

WE'RE NOT AS GOOD AS WE THINK WE ARE

Instead of seeing this as disloyalty, these observations opened our eyes to new possibilities

WE CAN LEARN FROM OTHERS AND CONTINUOUSLY IMPROVE

- We can incorporate outside practices into our own without giving up who we are or lowering the bar on the performance standards in our core work.
- Rather than disloyalty, it is our responsibility to lead change and improve our management practices as loyalty to our core purpose and our people.
- This awareness empowers senior managers to look more critically at our management practices and take action for the good of the organization and the business.

Thus, MOD's third annual retreat with Allen, in October 2006, was all about our new awareness and what it meant to us. Like the previous MOD management retreats, this was not intended to be a 'corporate vacation'. We met for two days away from the office and the distractions of our day jobs so that we could contend with everything we had seen and rally the entire management team around taking action. When we wrapped up, we had decided to change how we trained flight controllers and staffed Mission Control. This was near heresy because we were changing how we had done this core work since Project Mercury in the early 1960s, and were combining our operations training and flight control divisions as a result. Our decisions were heavily influenced by our benchmarking observations (especially from the 50th Space Wing) that demonstrated undeniable technical and financial benefits.

But the devil is in the details. We spent the next several months bogged down in arguments about how to take just the first step in changing how we had done our core work for more than 40 years.

Although we were still struggling to take specific next steps, something empowering had changed: rather than avoiding the ripples in the pond, MOD management was starting to see that the ripples were there whether we acknowledged them or not. Ignoring them and not talking about them didn't make the ripples and the problems that caused them go away. It just meant we were not choosing to deliberately manage the problems, which put us at risk of further decline, if not failure. We were now aware that it was *up to us* to do something about our management practices, or accept that it was *because of us* (if we did not do something) that the organization was allowed to decline.

▸ The hardest realization for leaders to make is that we are the problem, if we are not choosing to be the solution.

▸ Ignoring the ripples and the problems that cause them won't make them go away – we're just not choosing to deliberately manage the problems, which puts us at risk of further decline, if not failure.

▸ It is up to us to do something about our management practices, or it is because of us if we do not that the organization continues to decline.

Sea Change – Alignment

We were closing in on taking action but still missing some final step to get there. Although it was starting to look like MOD management had gone retreat-happy, we retreated again in April 2007 to finally do something! Before we did, our organizational strategy consultant, Deb Duarte, introduced Allen and his direct reports to John Kotter's book *Leading Change*.

After reading it, we gathered to compare notes as a group. This review was not much different from an elementary school book review, except we weren't a bunch of impressionable school kids. We were MOD's senior managers and a wary group of Mission Control leaders (I was now Allen's Deputy Director).

We compared the main takeaways we each saw throughout *Leading Change* and talked about how or whether they applied to MOD's management environment. To our unhappy surprise, we recognized our management behaviours in Kotter's description of eight common errors in making organizational change, for example:

allowing too much complacency and underestimating the power of vision. As we saw more examples of our behaviours in the common errors, we were also able to see the connection between our behaviours and our continuing failure to make real progress in changing anything. The hardest realization was that more of us saw *us* as the problem.

That realization, like our benchmarking, was a Eureka! moment. Benchmarking had increased our sensitivity to potential lessons from the outside world. *Leading Change* then took us to the next step into willingness to take action based on those lessons. We were MOD, after all, and our Mission Control indoctrination kicked in: We could not sit idly by if *we* were the problem; we couldn't allow ourselves to be the problem; we are always the solution, and if that wasn't the case, *we* had to do something about it.

Leading Change gave us a deliberate approach to taking action and increasing our chances of succeeding. Kotter lays out this approach by revising his eight common errors into a process for creating lasting change:

Eight-Stage Process of Creating Major Change
1. Establishing a sense of urgency.
2. Creating the guiding coalition.
3. Developing a vision and strategy.
4. Communicating the change vision.
5. Empowering broad-based action.
6. Generating short-term wins.
7. Consolidating gains and producing more change.
8. Anchoring new approaches in the culture.

John P. Kotter, Leading Change[34]

We used Kotter's process as a framework to conduct the retreat to ensure we'd have real action at last, not another plan to continue studying. We gave *Leading Change* to the rest of the management team a couple of weeks before the retreat. Our first step in the retreat had Deb Duarte leading the entire management team through a book review and group discussion, like she had with us senior managers already. The larger group had the same reaction to the book that Allen's direct reports had, and the same reactions that we had all had to our benchmarking conclusions: it was clear we needed to take action and make changes, and we were ready to make those changes in areas that mattered to us.

As the retreat went on, many of the direct reports led discussions about our core values, how we saw ourselves providing key roles in today's human spaceflight, and how we envisioned our critical roles in future operations. Each discussion confirmed the management team's commitment to preserving our core values, and in so doing reassured each of us that we could change core areas such as spaceflight training and operations without betraying our values. At the end of this two-day retreat, we committed to combining our training and operations teams, significantly modernizing our approach to training flight controllers, reorganizing major divisions to reflect these changes, and cost reductions of 30–50 per cent as a result of the changes. And it was effective immediately, with concrete actions mapped out to make the changes happen.

True to form, once MOD decided to take *Leading Change* to heart, we made a science out of addressing each of Kotter's eight stages. Allen briefed all of MOD on the strategy, the resulting reorganization and the longer-term goals. We then visited every division as their own Division Chiefs conducted 'town hall' discussions with their entire workforce and helped communicate the plans again. Within a month, we had briefed our stakeholders, made the organizational changes and begun the working-level efforts to shift our training methods. The steps we took for each stage are summarized in the following table:

Eight-Stage Process of Creating Major Change as Implemented in MOD Adapted from John P. Kotter, *Leading Change*	
Stages of the Process	**As First Implemented by MOD in 2007**
1. Establishing a sense of urgency	The Director was candid with his direct reports and the full management team about our negative image and need to innovate in some core areas and reduce cost.
2. Creating the guiding coalition	Benchmarking results and studying *Leading Change* as a group helped increase the direct reports' willingness to engage and strategize difficult change.
3. Developing a vision and strategy	The Director developed a strategy presentation and white paper based on the retreat discussions. This was reviewed across all direct reports.
4. Communicating the change vision	The Director briefed the entire organization. The direct reports then briefed their organizations again in town hall-like settings. Feedback was used to update the communications. The Director then briefed his senior management and stakeholders.
5. Empowering broad-based action	We implemented the organization changes immediately. The newly combined management teams were given significant leeway in recommending training changes that could produce flight controllers more efficiently.
6. Generating short-term wins	Divisions that produced the first formal training revisions were used as the models for others to follow. Subsequent divisions each tended to further improve the previous progress.
7. Consolidating gains and producing more change	As proposed changes were accepted at the top level, they were formalized immediately. Staffing plans and budgets were adjusted after each formal change. Managers at all levels began to stand out based on their ability to effect positive change.
8. Anchoring new approaches in the culture	This level of engagement became the norm for ongoing management dialogue, not just for retreats on specific challenges.

In a true masterstroke, Allen then solidified his 'guiding coalition', in Kotter's terms, by formulating the MOD Board of Directors (MOD BOD). This Board was simply all of his direct reports – his staff and all the Division Chiefs. These people were present in every top-management meeting already, so it wasn't clear to everyone at first how this changed anything. However, now there was an expressed intent that all strategic discussions would include the full MOD BOD, not just a subset of the managers behind closed doors.

We then leveraged that full participation to help preserve the momentum from the retreat and ensure that we took action. For example, in order to commit to the resulting cost savings these changes would yield, we first had to understand the current cost drivers within each division. Only then could we accurately discount our costs as we changed our training methods. However, although it sounds ridiculous in the retelling, we had spent so many years without insight below the executive level that we had to start from scratch to establish each division's bases-of-estimates (BOEs), connect them to the budget at the division level, and then trace them up to the directorate (executive) level.

Therefore, the MOD BOD's first formal action was to baseline a directorate level budget against a work-breakdown structure and then allocate it to each of the divisions. For the first time in years, the entire management team was privy to the MOD-wide budget and each division's BOEs.

The MOD BOD also collaborated on an organization-wide set of priorities. These priorities were intended to establish guidelines for all divisions. As the divisions were individually making changes in their operations training, work commitments and financial decisions these priorities set a common methodology for deciding what to keep and what to cancel.

At the end of his third year, it was a noticeably different management environment Allen Flynt turned over to me when I succeeded him as Director of Mission Operations. It had now all come together: where Allen couldn't convince the MOD management team that we had areas to improve, benchmarking showed us with our own eyes; and Kotter's *Leading Change* gave us a framework to better see our ongoing errors and then formulate effective change and take real action. And just as Allen had started out to do, this progress was taking root as a cultural expectation, not just by the executive but by all of the senior managers.

- ▸ John Kotter's *Leading Change* provides a framework to lead effectively and take action not only in change leadership but also in ongoing management.
- ▸ A key enabler in transforming the leadership team is found by reframing all of the executive's direct reports as a 'Board of Directors' with an expectation that all strategic discussions include the full Board, not just a subset of the managers.
- ▸ As confidence builds, this Board evolves into the role of Kotter's 'guiding coalition' in all decision-making, acting as leaders of the entire organization, not as representatives of individual business units.

Tidal Wave of Evolution – Transparency

We started 2008 with what had become an annual management retreat to set priorities and direction for the ensuing year. We still had detailed actions to plan and perform for the changes we had committed to at the end of the previous year, and we were sticking with Kotter's Eight-Stage process to do it.

Having seen the powerful influence that *Leading Change* had on the entire management team, the MOD BOD conducted another book review, this time Stephen M.R. Covey's aptly named *The Speed of Trust: The One Thing That Changes Everything*. It was an informal study again, with all of the direct reports reading the book in the week or two before the retreat. The goal was the same for this review: look for ideas that we could leverage to come together as a management team and improve our management practices.

When the MOD BOD gathered to compare notes on the significant takeaways the night before the retreat, it was clear that *The Speed of Trust* had had a profound impact on each of us. Once again, the initial impact wasn't a happy one. We found example after example of 'low-trust' behaviours that were still prevalent in our management ranks at all levels. Although we'd made considerable progress in the recent MOD BOD discussions, most of our management practices hadn't changed. We still struggled to have open discussion about our challenges and steps we could take to improve and find new opportunities. We still tended to operate like competing business units rather than as an integrated team or the Board of Directors we called ourselves.

Look again at the management practices at the end of Chapter 4, and compare them to the table below of low- and high-trust behaviours from *The Speed of Trust*. We reached three conclusions. First, we all believed that the high-trust behaviours described the organization we *thought* we were in – the one we *wanted* to be in – and demanded for our working-level people in the Mission Control Room. Secondly, of the high-trust behaviours listed, we could claim only 'a high degree of accountability', particularly when we were discussing our core technical work. Lastly, however, *all* of the low-trust behaviours were prevalent in our management ranks, including at the MOD BOD level. *We* were still the problem.

Cultural Behaviours in Low- and High-Trust Organizations From *The Speed of Trust: The One Thing That Changes Everything*, Stephen M. R. Covey[35]	
Low-Trust Organizations	**High-Trust Organizations**
People manipulate or distort facts	There is real communication and real collaboration
People withhold and hoard information	Information is shared openly

Cultural Behaviours in Low- and High-Trust Organizations From *The Speed of Trust: The One Thing That Changes Everything,* Stephen M. R. Covey[35]	
Low-Trust Organizations	**High-Trust Organizations**
Getting the credit is very important	People share credit abundantly
People spin the truth to their advantage	People are candid and authentic
New ideas are openly resisted and stifled	The culture is innovative and creative
Mistakes are covered up or covered over	Mistakes are tolerated and encouraged as a way of learning
Most people are involved in a blame game, bad-mouthing others	People are loyal to those who are absent
There are numerous 'meetings after the meetings'	There are few 'meetings after meetings'
There are many 'undiscussables'	Transparency is a practised value
People tend to over-promise and under-deliver	There is a high degree of accountability
There are a lot of violated expectations, for which people try to make excuses	
People pretend bad things aren't happening or are in denial	People talk straight and confront real issues
The energy level is low	There is a palpable vitality and energy – people can feel the positive momentum
People often feel unproductive tension – sometimes even fear	

We finished this MOD BOD 'book club' (as we'd come to call our book-review process) with a consensus that, as leaders, it was up to us to do something about our ongoing lack of trust within our own team. We immediately set out to do just that by looking for the management practices that reinforced low- instead of high-trust behaviours, and then change them.

▸ Stephen M.R. Covey's *The Speed of Trust* shows us the high-trust behaviours that may describe the organization we think we are in – the one we want to be in – however, it is our actual behaviours as leaders that determine the real trust level at all levels of the organization.

Expectations

I didn't have all of the answers, but like the rest of the MOD BOD, the problem now had my attention. I was listening for clues in our team discussions and in my one-on-one meetings with my direct reports: What kept us in the low-trust behaviours? What could we do differently to move to high trust?

I wasn't looking to solve it all with one flash of inspiration. Like tough technical problems, I was hoping to go one step – one low-trust behaviour – at a time. In meeting after meeting in our normal work, I'd poke at the subject of trust, often keeping the MOD BOD back after a larger meeting to pursue trust-related questions about some recommendation or report from one of the divisions: What are we missing in this discussion from other divisions' perspective? How did this problem come about or surprise us? Why hadn't a problem that applied more broadly in the organization been coord-inated with the other affected divisions? Why hadn't the MOD BOD heard this already?

It didn't take long before one of my direct reports spoke up. 'Nobody knows what you want. Do you trust us, or are you going to make all of the decisions?'

Some interpreted my focus on awareness at my level as micro-management. In response, we circulated a written policy to all divisions:

Coordination Policy	
As Applied in MOD	**Written to Apply Anywhere**
The following must be coordinated with the full MOD BOD:	The following must be coordinated with all of the senior executive's direct reports:
Decisions, positions and issues known or suspected to create concern among the senior program management, external organizations or center senior staff	Decisions, positions and issues known or suspected to create concern among our customers, stakeholders and senior management
Significant change to an existing operating philosophy, flight technique, procedure or process	Significant change to an existing operating philosophy or management practice
MOD resource impacts	Budget and personnel impacts
Non-standard processes, risks and decisions	Non-standard processes, risks and decisions

When I first distributed the policy for comment to the MOD BOD, I thought the whole thing was common sense and unnecessary bureaucracy. Wrong. I hadn't anticipated that the conversations we'd have about the full intent of each line of the policy would clarify the ultimate goal, not just to them, but also to me: the actual intent was *MOD BOD* awareness, not *Director* awareness alone. This wasn't about micromanaging each division from the executive level. Instead, it was about ensuring that *all divisions* were aware of *all the risks* we were managing across the organization. It was also to ensure all divisions were privy to, and able to benefit from, each other's successes and mis-steps rather than everyone reinventing the wheel. Yes, I would also be made aware, but more importantly, *they* were all kept informed and therefore better able to give me and their peers the benefit of their judgement.

Further, the value was not just in the *awareness*. The higher goal was the opportunity for the entire MOD BOD to *engage* in every subject that affected organizational strategy and risks. Where we once would have avoided the discussions for fear of putting ripples in the pond, we now focused on identifying the ripples as early as possible, and then managing them deliberately as a team.

As we discussed it, I made it clear to the team that the coordination policy did not apply only to *them*, but that they could hold *me* to the same expectations. What an easy price for me to pay as an executive: I shared with them all of the strategy- and risk-related decisions I was considering, and in return, each of my direct reports shared the same with the full MOD BOD. I had the greatest assurance of making informed decisions and setting good strategy – in leading well – because I had the senior leader of every MOD organization giving me guidance. More importantly, *we* had the greatest assurance of leading MOD well because the entire senior management team was engaged. That was the expectation. And it began directly addressing Covey's high-trust behaviours, such as real communication and collaboration, shared information, transparency and confronting real issues.

All of this became clearer to us, me included, not because we wrote it down, but because we discussed the underlying purpose at length. As we experienced the management transparency the coordination policy brought us, it became easier to identify and talk about the next barriers in our trust level and in our management practices.

- The real behaviour changes don't come simply from communicating the expectations but from open and fully engaged Board discussions about the underlying purpose.
- As transparency increases, it becomes easier to identify and talk about the next barriers in trust level and management practices as an integrated team.

Discipline

We then realized we needed to increase the rigour in our business practices. Although we had made considerable progress in developing and communicating a traceable financial baseline, maintaining an accurate awareness at all levels was proving impossible. The daily and weekly churn in managing the $650 million enterprise was a challenge to keep up with on its own. Add to that the significant operating and organizational changes we had started implementing in the past year, and it was a recipe for chaos. The divisions and the MOD business office were in constant disagreement over the effects of some operations change, any related resource savings, and the resulting division resource allocations.

That's when we discovered what most businesses already knew: for an ongoing enterprise, a baseline without configuration control is only a snapshot in time, and the accuracy quickly fades as business ebbs and flows. The MOD BOD accepted the pleas from the business office and gave them authority over a process that tracked the baseline and all proposed changes. As divisions found the need for more resources, they made their case to the MOD BOD. The business office assessed the impact to the overall MOD budget and tracked our performance through each budget cycle. If we moved resources between divisions, again, it was the business office that brokered the deal and tracked the decisions. Within a year, every division had instant access to a custom database of our financial baseline and all changes that were under consideration.

The business office also managed MOD-wide financial performance and was a secondary check to ensure divisions were bringing performance issues to the MOD BOD for discussions. Although this could have become adversarial, the business office's chief was a direct report, that is, a MOD BOD member. Our focus on high-trust behaviours and coordination policy – discussions she had been a part of – all helped diffuse those concerns. While we still had hard decisions to make, this process-driven rigour in our business processes added the discipline we needed to ensure we didn't lose the insight we had worked so hard to gain, as we continued changing and improving how we did business.

In an extension of this financial-baseline discipline, we added the same insight into our development projects. There was a continuous churn of new computer systems and software to support flight controllers and instructors. Rather than continuing to handle them in an ad hoc fashion, we required each division's project managers to periodically summarize the ongoing risks, cost, schedule and technical performance of every development project. The full MOD BOD was then able to debate the business case of any new project as well as any project that was underway. We focused on return-on-investment as well as top-level benefits according to the MOD BOD-approved priorities.

This bunch of rocket scientists was bringing business rigour to MOD. Rather than bogging us down in accounting (like so many technical managers fear), the greater insight increased the engagement and trust across the MOD BOD as we honed our new management practices.

▶ To increase trust across the Board, assess and modify every management practice for full transparency and engagement in every subject that affects organizational strategy and risk.

Inclusion

We had come far in MOD BOD engagement and our new business processes. However, I was disappointed to hear some of my direct reports complain about being left out of certain decisions. I heard this occasionally as we wrapped up a MOD BOD discussion and people were milling out of the conference room. I also heard it from time to time in one-on-one tag-ups with my direct reports each month.

I was surprised by two common themes: 'We all know you gave some of the management reserve to one of the divisions without talking to the rest of us' and 'We all know that you trust your immediate staff more than the rest of us'.

How the hell did they get those ideas? Replaying their comments in my head over and over, I realized that *I* was the problem.

First, I was still doing what Directors had always done – taking care of some easy and obvious problems behind closed doors. These weren't big, MOD-threatening issues. Every now and then a Division Chief would stop by with a concern over a small financial problem. Maybe they wanted $50,000 to buy some specialized software for their people or a similar amount to cover a minor cost overrun in a larger MOD BOD-approved project. No problem. MOD typically had several million dollars in management reserve for the year, and we could easily cover a small amount like $50,000 without impacting higher-priority work.

Why not make those small problems go away for each of my direct reports when I could, when the rest of the MOD BOD would surely have agreed? Because with our newly transparent business practices, the entire MOD BOD quickly discovered each change in resource allocations. They saw that I'd 'helped' one of their peers – that I'd made the decision without giving them the chance to evaluate the business case, either on its own merits or compared to similar problems we could have fixed for them. And it cost me the very trust that I was working to improve across the MOD BOD.

This was an easy behavioural problem to fix. Since I was the problem, I could be the solution. I told the MOD BOD at our next meeting that I still wanted to hear their concerns in our one-on-one tag-ups but that I would no longer make a final decision in private that was in conflict with our coordination policy. I'd help them strategize on how to fix the problem or how to work through it with another division, but any final decisions would first be aired in front of the MOD BOD.

▸ To avoid perpetuating low-trust behaviours, the executive must adhere to the same expectations for transparency and engagement as the team – give advice in private but make decisions with the full participation of the Board.

Secondly, I realized familiarity was helping the rapport with my immediate staff, and that was sending incorrect signals to the rest of the MOD BOD. Like many executives, several of my direct reports shared office space in a suite with me in our Headquarters building, while all of the Division Chiefs' offices were in other facilities with their divisions. The additional time I spent with my staff went a long way towards making them more comfortable telling me when I was wrong or not taking challenges from me personally. As a result, we did have an easier rapport in general.

My solution to this one was first to tell the MOD BOD that I hadn't realized this was a perception, and why I thought it was happening. My goal was to have a similar familiarity with each of the MOD BOD members and eliminate any appearance of favouritism or good-old-boy treatment. All of our future one-on-one tag-ups were scheduled in their facilities rather than my office, which kept them on their 'home turf', cut down on the number of times they were summoned to the boss's office, and got me into our working-level facilities much more often.

We also scheduled regular MOD BOD-only meetings, typically an hour every other Friday. The purpose was to discuss leadership, values and MOD culture. We worked hard to avoid discussion about normal work which would have turned these MOD BOD meetings into just more management meetings. Unlike many of our previous meetings, in which the MOD BOD was welcome to bring their deputies or to allow their deputies to represent them, these meetings were reserved for direct reports – MOD BOD only. We found that the presence of even a single deputy changed the behaviour of many of the MOD BOD, and they moderated their behaviour 'in front of witnesses' who hadn't gone through this evolution with us and might misinterpret the open and unfiltered discussion. Besides, I wasn't looking for each division to be 'represented' in our cultural discussions. I was looking for full and open engagement from the entire top tier of MOD leaders *as MOD leaders*, not as managers of some specific piece of the whole.

To my everlasting surprise and gratitude, these changes worked over time, better than I would have dreamed. Once again, across the MOD BOD, engagement and trust increased – the high-trust behaviours were taking root. As they did, it also became more natural for us to discuss concerns with our management practices, alignment with our values, culture etc. All it had cost me was the patience to wait for a MOD BOD discussion before final commitment (that is, applying our coordination policy to myself in all things) and an hour every other week with my most talented leaders dedicated to understanding the culture we all prized.

▶ Conduct regular, Board-only meetings to further normalize discussing the organization's leadership, values and culture and to reinforce alignment, transparency and engaged leadership.

Collaboration

As the MOD BOD became more comfortable with our newfound transparency and engagement, it became easier to collaborate on more things that had previously only been done 'privately' within each division. The MOD BOD had each become more comfortable in their role as a senior leader *for MOD* rather than just managers of their separate business units. That manifested itself next in personnel development.

For years, the senior managers (in positions that would later become the MOD BOD) had periodically reviewed a list of experienced personnel who were considered high-potentials for promotion in grade and into more senior-management positions. In keeping with 'no ripples in the pond', this typically amounted to each Division Chief defending the promotablity of 'their guy' and remaining silent about the other divisions' people. From 2005 to 2007, this had progressed to a more open exchange of comments and critiques of any of the high-potentials who were brought forward. By 2008, this had become real collaboration, with Division Chiefs making detailed evaluations of each other's high-potentials and following up with each other for additional clarification on areas that needed improvement. The results of these discussions were often reflected in the performance and development plans for each of their direct reports as all of the Division Chiefs benefited from the perspective of peers who might offer observations that weren't obvious closer to home.

2008 was the second year of the MOD BOD's greater awareness of our challenges and the shortcomings in our management practices, and more importantly, it was the second year of our increasing willingness to admit them out loud. In those two years, *Leading Change* and *The Speed of Trust* had got our attention and shown us how to say some of the things many of us understood, but struggled to put into words. As the MOD BOD wrestled with each next weakness or challenge, we made one change after another in our management practices, as shown in the table below:

MOD Management Practices Implemented as a Result of Applying *Leading Change* and *The Speed of Trust*
Established the MOD Board of Directors
MOD BOD baselined a comprehensive budget against a work-breakdown strategy
MOD BOD-only retreats to set priorities and strategies
Formal policy established to coordinate all issues, operating philosophy and resource changes, and non-standard decisions
MOD BOD established top-down configuration control in business processes
Quarterly MOD BOD review of all development projects for ROI and ongoing cost, schedule and technical performance
No more decisions in 1-on-1 meetings between the executive and MOD BOD members
Biweekly 1- to 2-hour MOD BOD meetings, focused on leadership, values and MOD culture
MOD BOD collaboration on succession management

Each change was another step towards high-trust behaviours, and each decision further strengthened the MOD BOD's alignment to our common cause. We regularly applied *Leading Change*'s methodology to 'keep us honest' in our implementation, and we applied *The Speed of Trust*'s high trust as our barometer for how we were doing and where else we could improve.

In this year, not only did we implement this raft of management-practice changes, but the increased trust level became more obvious around the conference table in every MOD BOD meeting. We had learned that how we managed mattered and was determined by how we engaged with each other.

It was working. The MOD BOD was moving towards high-trust behaviours. Trust was increasing among them, and between them and me.

We had started to undo the effects of the management cloud!

But it was still tenuous. Real and lasting trust takes time, especially when the goal is undoing long-held practices and evolving cultural norms. So the MOD BOD continued the discussions and introspection.

The New Normal – Stewardship

By the end of 2008, the deliberate transparency and the more collaborative management environment it created had not only become more natural to the MOD BOD, but were now also expectations. Of course, I expected those behaviours from them because I saw how powerful they were in our ongoing management; however, the *MOD BOD* now expected it from *each other* and from *me*. They had bought in and were now engaging with each other as *leaders of the enterprise*, not just managers of their individual areas of responsibility. The MOD BOD had become an aligned leadership team.

Although we'd come a long way in this year, a few of the Division Chiefs still fell back into old habits from time to time, particularly advocating for 'their people' or 'their division's' pet projects, instead of decisions that made sense for MOD as a whole. Even so, they all assured me they understood what it meant to be *leaders for MOD*, not just parochial managers 'protecting' what their divisions had.

My deputy, Steve Koerner, and I took them up on it. We rotated a few of them between divisions. As we strategized on the best moves for the individual leaders and for the organization, Steve realized that those who still had the greatest tendency to circle the wagons were managing the same division they had been in for their entire NASA career.

Bingo!

At least some part of their behaviour could be attributed to seeing the world through only that lens as they progressed up into the management cloud, and perhaps a change would help them see that for themselves.

This rotation was a bitter pill for a couple of them to swallow. Our senior managers typically were promoted from within the same organization and naturally wanted nothing more than to lead their home organization. They saw their rotation as punishment of some kind, maybe for not agreeing with me on some strategic decision. But no, our goal was to bring some additional awareness to these talented leaders by moving them into a leadership role that was outside of the area in which they were the technical experts. That way, they'd be forced to take the higher view, leave more of the detailed, down-and-in technical discussions to their management team, and engage more openly with the MOD BOD.

This move set us back for a while in the overall trust level around the MOD BOD, as many wondered who was next in the musical chairs. Over the next six months or so, all but one of them came round and saw the benefit – to the individual leaders and to each of the divisions – of having a fresh perspective. They also realized that the rotated leaders hadn't been banished from the island. They were simply transferred to other critical leadership positions and still responsible for key divisions. This wasn't about punishment, it was about increasing the effectiveness of critical leaders in critical leadership roles. The rotations also succeeded in reducing the 'wagon circling' from the rotated Division Chiefs (again, except for one). Within a year, we had some of their peers offering to rotate to a different leadership position for the same purposes.

▸ Rotating leaders into leadership roles outside of the area in which they are experts helps them take the higher view, leave more of the detailed, down-and-in discussions to their team, and engage and collaborate more openly with their peers.

We started 2009 with our annual management-team retreat. Instead of current challenges and tactical actions, we shifted the overall retreat focus to match the MOD BOD's more closely: top priorities, strategic challenges, and leadership culture. Thanks to the work Deb Duarte had done with us for several years, and the progress we had made by applying Kotter's and Covey's insights, we were able to facilitate this retreat (and all that followed) for ourselves. My role was typically only influencing the agenda, kicking the meeting off and summarizing conclusions and next steps at the end. Working-group sessions and any larger group discussions were led by some other member of the MOD BOD, *acting on behalf of the MOD BOD*, not just of their division. Each time we did it, like the high-trust behaviours, it became easier and more natural.

This retreat started with a MOD BOD book-club discussion of Jim Collins's *Good to Great*. Our *Leading Change* and *The Speed of Trust* book-club experiences played out yet again. As we read explanations for some of the management behaviours we were trying to move away from, we had a number of 'a-ha! moments'. These were more insights into attitudes many of us had but couldn't yet articulate, for example[36]:

- The moment you feel the need to tightly manage someone, you've made a hiring mistake. The best people don't need to be managed. Guided, taught – yes. But not tightly managed.
- Put your best people on your biggest opportunities, not your biggest problems.
- Yes, leadership is about vision. But leadership is equally about creating a climate where the truth is heard and the brutal facts confronted.
- The Stockdale Paradox: Retain faith that you will prevail in the end, regardless of the difficulties, *AND at the same time* confront the most brutal facts of your current reality, what-ever they might be. (Named for former prisoner of war and Congressional Medal of Honor awardee Vice Admiral James Stockdale.)

Good to Great's lessons reinforced many of the management practices we had put in place in the previous two years. The next year again found these practices to be easier, more natural, and considered to be the norm by all levels of our management team. The boost *Good to Great* had given us in articulating the value of these practices was also important in preserving the MOD BOD's focus on high-trust behaviours and in managing a variety of strategic-change efforts down and in the organization.

....

It Isn't a Core Competency If We Don't Know How We Do It

To kick off our retreat, we invited the ISS Program Manager to share his thoughts about MOD's recent accomplishments. This was partially intended to be a feel-good opportunity, because this program manager was very happy with us that year. In the previous year, we had given his program a permanent 23 per cent reduction in our ISS flight operations labour costs – and they had not been expecting the rebate. As expected, he regaled our management team with his high regard for not only our technical performance but now also our business prowess.

As this important customer told us how great we were, I watched all members of MOD management nodding their heads and smiling with pride. 'You're darn right we're good' was written all over their faces.

Afterwards, in what was intended to be another feel-good and rah-rah exercise, the entire team was discussing MOD's core competencies. As expected, one after another claimed leadership and leadership development as some of our top core competencies. And why not? We were known for it agency-wide, as my dad had pointed out to me years before. We'd also just heard one of our most important customers confirm it.

As I listened, something began nagging at me. I stopped the group discussion and asked all 60 or so of these MOD managers, 'Why do we think leadership is a core competency? I hear us all saying it. I know that leadership is important to us, and that MOD has a history of producing strong leaders. But does that make it a competency?'

Yikes. I thought I was going to have a mutiny on my hands. The room erupted, with some even jumping to their feet and raising their voices. Among other things, I heard, 'Are you saying MOD doesn't produce great leaders?'

I answered, 'I didn't say we don't produce leaders. But just doing something doesn't make it a competency. Maybe we've just been lucky in employing people with natural leadership ability. Maybe our job is so tough – so demanding of strong leadership – that it produces good leaders organically, and all we do is cherry-pick them.'

A respected middle-level manager who reported to a MOD BOD member was on his feet and visibly angry. 'How can you say MOD doesn't produce leaders?' he asked.

I said, 'You're not hearing me. Look at it another way. If it's a competency, surely you're grooming it into your people. Can you tell me three things – no *one* thing – that *you* did in the previous year to *intentionally* enhance the leadership potential for one person who reports to you?'

He stood there glaring at me. 'I may not be able to say how we do it, but we all know that MOD produces leaders. I know it when I see it, even if I can't say it.' 'That's my point, folks. Knowing it when we see it and producing leaders unintentionally doesn't make it a core competency, no matter how important it is to us. We are MOD. Since when do we do anything unintentionally? We'd never tolerate that in our technical work. But think about what it means if MOD decides to be as deliberate in developing leaders as we are in flight operations. You think we've produced leaders in the past? If MOD does it on purpose, think of the leaders we'd produce!' As I said this, half of the room still looked bewildered. The other half understood my point.

Most of the MOD BOD looked a little like a deer in the headlights with a dawning realization. After all of our progress in discussing values, leadership and culture, evolving our management practices to high-trust behaviours, and impressing this normally unimpressible customer, none of us could say how we produced leaders, or even what a leader looked like when we did.

'I'm not saying I have the answers, folks. But we're not going to let this go. If this bothers you, good. It bothers me too, because this *is* important to us. Keep thinking about it. More to come . . .'

....

In our next biweekly MOD BOD meeting about leadership and culture, we picked up where that retreat discussion had ended. I spent 30 minutes summarizing my view of the world – much of it heavily influenced by our last couple of years' journey through acknowledging our weak areas, aligned change leadership and high-trust behaviours. I also talked about MOD's strengths and core purpose. When I finished, I said, 'Look, this is just how I see it. That doesn't mean I'm right. I'm not looking for you to parrot my words. I want to hear what you think. Where am I wrong? What did I leave out? How could we say it better – more accurately?

'Starting at our next MOD BOD meeting, in two weeks, each of you will have your chance to stand up and tell the rest of us your view of the world. How do you describe MOD's leadership values and culture? You all just heard me, so how hard can it be, right?'

First up was the toughest talking of the bunch, Chief Flight Director John McCullough. As usual, he was eloquent and passionate, and he described our leadership values and culture better than I had. He started by saying, 'When Paul first asked us to do this, I thought it was a waste of time. We all know this stuff. Why spend time saying it? When I started writing my notes, I didn't know how to start, even though I thought it would take me only a few minutes. Writing this down was one of the hardest things I've ever done.'

And over the next six months or so, every one of the MOD BOD got up in front of this small aligned group of leaders (and peers). They each nervously agreed with how difficult it was to articulate, even after hearing not just me but all of their peers who had already spoken. Each experience was like our book clubs. We heard some new comment or perspective that hadn't yet been mentioned, another 'a-ha! moment', where we all then compared notes on what that really meant . . . how we really see and say that.

By the end of the year, every one of us could answer those questions about our leadership values, how they set MOD apart, and how they made us capable of being so successful at the work that could not be allowed to fail. We didn't speak from a script, and we didn't simply parrot what we had learned from *Leading Change*, *The Speed of Trust* and *Good to Great*. However, we each applied what we had learned from the book clubs and from our regular leadership, value and culture discussions. Each member of the MOD BOD used slightly different words and their own style and passion, but each was able to articulate the values, not just to each other, but also down the line to the next level of the management team.

This experience led to our ability to articulate the real-time morality more clearly than ever before as it applied in the Mission Control Room, after almost 50 years of flight operations! In fact, it was this exercise that ultimately led to the description of the real-time morality, as written in Chapter 3, and our understanding of its role as a key enabler.

Now that we could all see it and say it more clearly, we could all see the direct connection between our specific management practices and our leadership values. Once again, seeing that connection reinforced the management practices we'd put in place since 2006, and further rooted them as expectations from the entire MOD BOD.

In 2009, we took collaboration on high-potential emerging leaders to the next level. The management-succession discussions had gone from informally influencing development to dominating the MOD BOD-wide perspective on all of the high-potentials. When I promoted someone into a MOD BOD position, I first reviewed the candidate list with the full MOD BOD and listened to their thoughts about each candidate's readiness to manage in our transparent and engaged – high-trust – environment. Interviews focused on a candidate's readiness to grow into a high-trust management style, as much or more than they did on domain knowledge and management experience. Afterwards, Steve Koerner and I explained final selections to the MOD BOD in the same terms. In turn, each MOD BOD member did the same when they were selecting a direct report. They first compared notes with the MOD BOD about the candidates, looked for readiness to evolve into high-trust management, and summarized their selections to the rest of us accordingly.

▸ Selecting leaders with a predisposition for the high-trust management style increases the effectiveness of the cultural evolution and the trust at all levels of the organization.

We ended this year having increased the trust level around the conference table in the MOD BOD to a very high and near-organic level. During one meeting, as we discussed some contentious topic, I was amazed at how far we had come, and I had my own 'a-ha! moment'. We had stepped up the role of the MOD BOD beyond just transparency and engagement, and I realized that I had pushed my authority down to the MOD BOD. Rather than costing me authority, it had made us a better leadership team. Every decision and strategy was better with this larger group of leaders' influence, and they were much more willing to provide it without prompting.

This leadership team had embraced the evolution. It felt like we'd accomplished what Allen Flynt had started, and had made it a real, cultural change.

A month after this realization, the Constellation Program was cancelled. As I explained in Chapter 5, NASA descended into chaos. MOD was not immune, since this eliminated one of two major customers, half of our budget and half our workforce. In Chapter 9, we'll go into the critical risks we focused on and how our high-trust management practices served us as we managed through them. For now, I'll describe just one facet of our response, which followed our new pattern. After our first few weeks of immediate damage control, the MOD BOD retreated by ourselves to have some unfiltered, soul-searching discussions to ensure that we set strategies that mattered. And by 'mattered' we meant as defined by our core purpose and values, not as isolated business units.

Good to Great had proven so powerful in reinforcing the connection between our management practices and our values that we turned again to *Built to Last: Successful Habits of Visionary Companies* by Jim Collins and Jerry I. Porras, for a MOD BOD book-club discussion before we took on the tough decisions. *Built to Last* didn't let us down. Even in this crisis, we found more 'a-ha! moments' and great takeaways like[37]:

- Distinguish core values and enduring purpose from operating practices and business strategies.
- Stick to the knitting.

- The *knitting* is the core *ideology.*
- You do not 'create' or 'set' core ideology. You *discover* core ideology. It is not derived by looking to the external environment; you get at it by *looking inside.*
- *Ideological* control preserves the core, while *operational* autonomy stimulates progress.
- Cult-like tightness around an ideology actually *enables* a company to turn people loose to experiment, change, adapt and – above all – *act.*
- It's not the content of the ideology that makes a company visionary, it's the authenticity, discipline and consistency with which the ideology is lived – the degree of alignment – that differentiates visionary companies from the rest of the pack.
- [It] is not the quality of leadership that most separates the visionary companies from the comparison companies. [It] is the continuity of quality leadership that matters – continuity that preserves the core.
- Good enough never is.
- Beware the 'we've arrived syndrome' – complacent lethargy that arises once an organization has achieved a BHAG [Big Hairy Audacious Goal] and fails to replace it with another.
- Never forget to preserve the core while stimulating evolutionary progress.
- No room for the unwilling.

In addition to helping keep us focused on values and core purpose while we responded to crisis, we once again found the words to articulate ideas many of us were struggling with. This reinforced our existing management practices and led to several of our remaining significant changes in this evolution.

'Ruthless continuous improvement' and 'no room for the unwilling' may not sound like management practices, but they are certainly attitudes the MOD BOD embraced more consciously after this book club. Backed up by accepting the notion that 'we've never arrived', our transparent discussions became even more so as we continued to second-guess our decision-making and consistency to our values and the MOD BOD priorities.

Similarly, there was less and less patience from across the MOD BOD if any of their peers were perceived as circling the wagons for 'their division' or shying away from making the tough-but-correct decisions for the good of the organization and our core purpose. The goal wasn't an expectation to agree with every decision without dissent, but that every member of the MOD BOD was expected to be a willing participant in transparent and engaged leadership aligned to MOD's core purpose and value.

▸ Every member of the leadership team must be a willing participant in transparent and engaged leadership aligned to the organization's core purpose and values, but everything else is up for scrutiny and change.

Implemented as a result of applying *Good to Great* and *Built to Last*
Periodic 'surgical' MOD BOD rotations to catalyse values-driven leadership individually
Self-facilitated, annual leadership retreats on priorities, challenges, leadership culture
Select leaders for predisposition to values, behaviours – clock building
Push authority down, bring MOD BOD up/in on all decision-making
Ruthless continuous improvement
No room for the unwilling

We also began pushing down the MOD BOD evolution as a deliberate effort to introduce these awarenesses to the next level of managers and the emerging leaders who would replace us. Every manager who reported to a MOD BOD member was given a set of the books we had reviewed in book clubs: *Leading Change,*

The Speed of Trust, Good to Great and *Built to Last*. They were given a year to read them and participate in a book club with their peers and the MOD BOD member they reported to. The goal was the same as the MOD BOD's – compare the takeaways to what they find in their division and across MOD. The dialogue was the goal, not the specific quotes that resonated with them. It had the same effect on each of the divisions' management teams as it had had on the MOD BOD. Not only did they learn to articulate ideas they believed but hadn't put into words, but they also saw the open dialogue demonstrated by their bosses about those ideas and gained confidence in the values focus for themselves.

We then began to rotate the MOD BOD's direct reports across division lines, for the same developmental reasons we had rotated some of the MOD BOD. Once again, the change in perspective was almost always a catalyst for increased leadership effectiveness. The individual and the organization both benefited.

In this same time frame, the MOD BOD's high-trust way of doing business had become such a normal part of our experience that the MOD BOD's deputies had seen and heard the effects for themselves. They wanted in and began asking to participate in our normally closed-door sessions about leadership values and culture. I originally turned down their requests because of the previous concern that their presence would weaken the MOD BOD engagement. However, the rest of the MOD BOD disagreed because they believed our trust level had become so high. From then on, the deputies participated in most MOD BOD sessions. This gave us more frequent opportunities to demonstrate our transparency and engagement with the leaders who were most likely destined to replace us over time.

The MOD BOD had come to grips with our leadership values and the management practices that reflected and reinforced them. Those values then dominated all of our decision-making. The resulting management practices became the expected norms of (and from) the leadership team. We had taken deliberate steps to develop those leadership values and behaviours into the next generation of leaders, and had shifted from exploring these ideas and experimenting with management practices to stewarding the ideas as the foundation for our leadership culture.

This transparency and our values focus had put us on our way to leading with the same technical truth, integrity and courage that we relied on in the Mission Control Room. We were bringing the real-time morality into our most senior management ranks and undoing the effects of the management cloud! Our jobs may have become more 'nuanced' in the management cloud, as we saw in Part 2, but the morality did not. We may use different words in our management roles – transparency, values alignment and engagement – but the effects are the same in all decision-making: putting all cards on the table (technical truth); deliberately discussing and applying our core values (integrity); and speaking up in spite of the intimidation (courage). Just like in real-time, each of the trust elements strengthens the next.

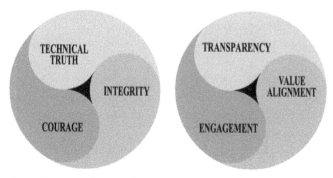

Trust elements of the real-time morality
At the working level (left) and in management (right)

Not bad for a bunch of rocket scientists and engineers, most of whom are not nearly as comfortable talking about the 'touchy-feely stuff ' as we are solving technical problems!

In Chapter 10, we'll take a more specific look at our leadership evolution and its application in other organizations, from the growth in awareness to the book-club process and management practices. The focus will be on duplicating the evolution with the benefit of a road map, as well as hindsight from our experience. Before we do, we turn to Chapter 8 and another critical contribution that is woven throughout the journey.

- The ideas from *Leading Change*, *The Speed of Trust*, *Good to Great* and *Built to Last* and the 'book club' discussions about them help normalize a values focus as a leadership team takes on more complex change leadership and new challenges.
- They show the way to bringing the real-time morality into the management ranks, undoing the effects of the management cloud and stewarding a leadership evolution already underway into a cultural evolution.
- In MOD's experience, benchmarking, *Leading Change* and *The Speed of Trust* got our attention, and showed us how to say some of the things many of us understood but struggled to put into words.
  - Each step in our series of management-practice changes was another step towards high-trust behaviours, and each decision further strengthened the MOD Board of Directors' alignment to our common cause.
  - We regularly applied *Leading Change*'s methodology to 'keep us honest' in our implementation, and we applied *The Speed of Trust's* high-trust behaviours as our barometer for how we were doing and where else we could improve.
  - We learned that how we managed mattered, and was determined by how we engaged with each other.
- *Good to Great* and *Built to Last* took us further into understanding the importance of aligning our management practices to our values and being able to discuss it in those terms.
  - Now that we could all see it and say it more clearly, we could all see the direct connection between our specific management practices and our leadership values.

- ▷ We were not only a more open and engaged leadership team, but we had become a more effective and aligned one as well.
- ▶ The leadership team embraced the evolution and engaged with each other as leaders of the enterprise, not just managers of their individual areas of responsibility.
  - ▷ The norm in all decision-making was now dominated by the real-time morality's unyielding alignment to purpose and deliberately applied high trust.
  - ▷ By demanding this in actual management practices, we were undoing the effects of the management cloud.
  - ▷ The real-time morality had become a cultural norm in our leadership team.

CHAPTER 8

Make It Personal, Make It Permanent

One of the greatest mistakes of successful people is the assumption, 'I am successful. I behave this way. Therefore, I must be successful because I behave this way!' The challenge is to make them see that sometimes they are successful in spite of this behavior.

Marshall Goldsmith, *What Got You Here Won't Get You There*[38]

Is It Me? – Awareness

The MOD BOD's leadership evolution started with the awareness that our challenges were not that different from other organizations', we didn't compare as favourably as we thought, and we could learn from those comparisons. With that awareness, we were able to grapple with the practices that were getting in our way and take deliberate action to become a very engaged and high-trust leadership team.

For us to evolve as a team, however, each of us first had to buy in to the high-trust management style, requiring each of us to evolve *individually*, or choose not to. Just like the MOD leadership team, many of us needed help gaining an individual awareness of our individual challenges and weak areas. While we are frequently aware of our individual strengths, we are often less aware of our weak areas, and how our strengths and weaknesses compare to those of the people around us. As a result, we don't see the clues that could point out our own individual behaviours that are getting in the way of our intended outcomes, just as MOD's management practices were getting in the way of the team's. And just like the MOD BOD, until we expand this personal awareness, it is difficult to evolve personally and buy in.

For many years, the Johnson Space Center used various forms of 360 feedback to help with this awareness. It's a great technique through which managers, peers, customers and suppliers offer individual observations, criticisms and suggestions. This feedback can be useful in understanding strengths, weaknesses, and how we are perceived versus our intentions. Unfortunately, for it to be most effective, we each need some level of this individual awareness already. Otherwise we run the risk of ignoring feedback that doesn't match our existing self-image, much as MOD management did not initially accept our less flattering benchmarking observations as relevant.

Compounding that challenge, the effects of the management cloud tended to make us less willing to talk about, or seriously consider, what were often the 'human element' problems described in Chapter 5. For example, as a Flight Director, I had years of receiving anonymous and unfiltered feedback from every team of flight controllers I led. Coming from the habitually direct communicators in Mission Control, most of this feedback left little room for interpretation in a Flight Director's leadership performance, both strengths and weaknesses. When delivering the news, our boss would often remind us, 'Pay most attention to the good stuff. Don't just ignore the criticisms, but you know your strengths. You know what got you here. Keep dancing with the one that brung ya.' Although that may have helped our egos, it certainly didn't encourage us to think about areas to improve.

▸ For a leadership team to evolve to a high-trust management style, each team member must first buy in, requiring leaders to evolve individually.

▸ We are frequently aware of our individual strengths; we are often less aware of our weak areas and how our strengths and weaknesses compare to those of the people around us.

▸ For feedback to be most effective, we each need some level of this individual awareness, otherwise we run the risk of ignoring feedback that doesn't match our existing self-image.

....

Sometimes Feedback Gets Your Attention

In February 2003, I had been a Shuttle and ISS Flight Director for seven years. I had files of flight controller feedback rating my talents as a Flight Director, 'sharp edges' and all. In the very specific context of leading the team in Mission Control, I had a very clear understanding of which of my strengths were most helpful to the teams, and I had a good idea of which of my tendencies they liked the least.

Overall, I had a reputation for getting things done, leading teams to solve hard problems, and being a hard-ass (which I saw as candour and a low tolerance for the unprepared or anything less than best effort).

This is when I was assigned to lead center-wide engineering teams in solving a wide range of 'impossible' problems revealed by the Space Shuttle Columbia accident. If we failed, it was feared that NASA wouldn't fly another Shuttle. Worse, these problems had been studied more than two decades earlier and formally declared 'impossible'. Still worse, the next Shuttle flight was scheduled to launch in six months, and we were expected to have solutions ready in time for it.

The pressure was on, and it was excruciating. Eclipsing the technical challenge, we had lost seven of our friends on Columbia who had counted on us to protect them. We couldn't fail them again. We couldn't let their deaths be in vain. It was my job to lead the team in finding solutions and protecting the astronauts who would fly next. I was determined not to let them down.

We went to work as only NASA can in response to crisis and challenge. Within four months, we'd found solutions that rendered the previously 'impossible' problems into only 'really hard' problems. Things were moving fast, and some parts of our community were not convinced that our solutions would work. We responded to one challenge after another with specific engineering tests and analyses designed to answer each concern. As we acquired more data, we solidified the recommendations. After leading the initial engineering investigations, I was then assigned to lead the Mission Control team in planning and flying the mission.

Although it took almost two more years to have all of the new equipment ready to fly, our team's success was critical to NASA resuming Shuttle flight operations in 2005. Despite scepticism from some parts of our community, all of the new equipment and operations were tested in orbit during STS-114 and worked as designed. We had demonstrated that the 'impossible' challenges had been solved after all, and Shuttle went back to work finishing the International Space Station.

Along the way, some parts of our community remained sceptical and criticized my leadership harshly to executives. It was not unusual to hear secondhand comments like, 'We know he's in cahoots with the management to pull some things over on the community so we can fly sooner, *even if it gets the next crew killed.*' That was soul-crushing.

The personal comments ate at me. However, I could handle public debate over engineering, so I focused on the engineering data and did all of my talking in open forums with managers and engineers present to reach their own conclusions. *I may not be able to convince them of my intentions,* I thought, *but I could damn well show them the facts.* I was definitely dancing with the one that brung me, and knew that the data and success were all that mattered when it was time to put astronauts back into harm's way.

By the time we flew Shuttle again, I had leaned on every professional experience I'd ever had to lead this team across so many engineering disciplines while solving these difficult problems. I was convinced there was no way I would ever have a bigger impact on my profession than succeeding in this leadership role. Reinforcing that conviction, I picked up several prestigious awards along the way.

After the flight, the Center Director (at the time, three levels of management above me) pulled me aside. 'Paul, you did a great job leading the return-to-flight for MOD. But you pissed off some influential people while doing it. It's time to get your resumé up to date for your post-NASA career.'

....

With the feedback that I had become persona non grata after the most important leadership role of my career, I had certainly become aware that, to some people, something else mattered besides the data and success. Gaining that single awareness wasn't enough.

However, while I was waiting for my NASA exit, I was sent on a management-development program. This was the kind of program Flight Directors typically criticized as 'psychobabble and a couple of days of good content crammed into two weeks'. In preparation for the program, I completed a Birkman behavioural assessment. Like other, similar assessments (for example, Myers-Briggs Type Indicator and DiSC) the Birkman is intended to help the participant understand their behaviours and communication styles. This was just the kind of thing that Flight Directors were warned to take with a grain of salt while we focused on our strengths.

After completing the lengthy assessment, a debriefing of the results with an executive coach changed my thinking on the value of the exercise. As the coach explained one result after another, and speculated on the situations that caused me the most stress, it was like she was reading my mind. I told her, 'There are facets of this that are such deeply buried personal feelings that even my wife doesn't really understand them about me. And yet, you're a complete stranger, and you've just read them to me in our first conversation!'

That led to more detailed sessions with another executive coach to further explore the implications of my survey compared to others'. That coach, Tim Andrews, helped me understand how people with different personalities interpreted my normal communication style, and why I responded to their communication styles the way I did. Looking back on my return-to-flight experience, it became clearer to me that some of the scepticism and lingering bad feelings were related to personality type and style conflicts, with little of the intended message ever having got through. Like my first Birkman debriefing, it was mind blowing to hear Tim accurately predict my own reactions and those soul-crushing feelings! Hearing my surprise, Tim said:

> Paul, 75 per cent of the people in the world don't even know how to say 'Hello' to you. But you're not alone. Even though most people in the world are built differently than you, we all have our own strengths, weaknesses and communication styles that can get in the way of our intended message. It's often hard for us to see the weaknesses for ourselves. Sometimes they're blind spots that we can't see at all. But we can become aware of them and choose to do something about them, rather than let them define us.

Through several more discussions, Tim helped me understand more specifically how these personality and communication-style differences led to assumptions about each other's intent – assumptions that were just as likely wrong as right and got in the way of credibility and trust. As I came to appreciate this, I also saw my own weak areas more clearly. I was then able to accept that many of the personality traits and communication styles that I was weakest in were also strengths I was lacking – that *my* strengths weren't the only strengths.

INDIVIDUAL AWARENESS GAINED FROM PERSONALITY ASSESSMENTS AND FEEDBACK

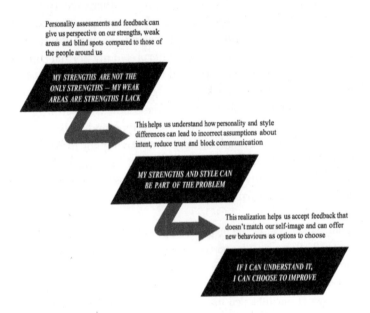

Personality assessments and feedback can give us perspective on our strengths, weak areas and blind spots compared to those of the people around us

MY STRENGTHS ARE NOT THE ONLY STRENGTHS – MY WEAK AREAS ARE STRENGTHS I LACK

This helps us understand how personality and style differences can lead to incorrect assumptions about intent, reduce trust and block communication

MY STRENGTHS AND STYLE CAN BE PART OF THE PROBLEM

This realization helps us accept feedback that doesn't match our self-image and can offer new behaviours as options to choose

IF I CAN UNDERSTAND IT, I CAN CHOOSE TO IMPROVE

After the difficult ordeal and great success of my teams during Shuttle's return-to-flight, it was easy to ignore any criticism of my leadership, even being told I was persona non grata. However, the Birkman assessment got my attention with its insightful descriptions of not just my personality and communication style but also my reactions when confronted by different styles and other people's reactions to me. That made me more willing to consider the Birkman insights into my own contribution to some of the conflict, intentional or not. Tim Andrews' executive coaching was critical in my willingness to embrace this new awareness about personality types, communication styles and my own tendencies that contributed to some unintended and sometimes avoidable 'bodies in my wake' as I led teams, including during Shuttle's return-to-flight.

With that awareness, I had an opportunity to do something about it.

▶ Personality-type assessments are invaluable in understanding our own tendencies, how our differences influence our interpretation of each other's intentions, and how the misinterpretation can affect behaviour and inhibit communication.

Intentions and Actions – Alignment

When I returned from the management-development program, I was still struggling with what to do about my new awareness. I could not get past the implication that, if I needed to change, there must be something wrong with me, or that I was the bad guy. I couldn't reconcile those thoughts with my reputation for leading teams to solve tough problems. I had a great track record of getting the mission done and protecting our astronauts, which was exactly my intention. How could this be on me? Why was I the one who needed to change?

A month later, Allen Flynt intervened in my departure from NASA and brought me back to MOD to manage all of our Shuttle operations. When he did, he also turned me over to his executive leadership coach, Lee Hayward, both to help me get unstuck on what to do with my new awareness, and to help me make a successful transition from strong technical leader to effective senior manager.

In our first conversation, I shared my recent experiences of becoming persona non grata, gaining awareness from the Birkman Assessment and Tim Andrews' coaching, and my goal to improve. As I talked, Lee stopped me from time to time to ask things like, 'Can you explain what you mean by that just a little more? How did you respond in that situation? Did it help? How do you think the person you were trying to influence heard it? Could you have responded differently and still gotten the results you needed?'

As I wrapped it up, I summarized by saying I apparently now needed to be 'fixed' in order to be a manager. Lee stopped me and said, 'I don't think you understand. I don't do "remedial" coaching. I'm not trying to fix anything. I don't think you're broken. I don't think Allen thinks you're broken either, or he wouldn't have promoted you. What I hear you saying is that you want to be more mindful of how you come across to other people, so your words – how you are received by others – match your intentions. I think you can do that, but it isn't about fixing yourself. It is about helping you be even more effective than you already are.'

Lee then surveyed my peers and others within our organization for feedback about their experiences with me. He also interviewed a number of people in manager, customer and supplier roles. The feedback and interviews were all along the lines of these sentence-completion exercises:

- When I think of Paul Hill as a leader, I think of him as _____
- The way I know Paul respects what I bring to the table is when _____
- Things would be better in our communication if Paul and I would _____
- Things would be better in our working relationship if Paul would _____
- Paul's greatest strength in his current role is _____
- Paul's inability to _____ often gets him in trouble by _____

The interview responses were particularly enlightening because Lee would follow up with the interviewee just as he did when he and I talked privately. He asked more questions for clarification, and they'd frequently come back with more detailed comments. When he reviewed the integrated feedback with me (without identifying who had made each comment) I was anticipating a long list of critical (if not negative) comments. Besides the still recent experience of being told I was persona non grata, I had recommended Lee interview a number of people who I knew to be critical of me. I was surprised by what he found. Although there were some critical comments and suggestions, they weren't really negative – they didn't suggest I was persona non grata or 'broken' after all. Most shocking to me was the overall positive theme and the specific comments citing strength after strength, especially since I had already accepted that I was considered a 'dead man walking'.

With so much positive feedback, I finally realized this wasn't about changing who I was, compromising my core values, or accepting that I was a bad guy in spite of my intentions. With that realization, I cleared my final mental hurdle in taking constructive action with my new awareness: I could improve and still be authentic – still be me. My formal coaching objective then became: learn to be more deliberate in my personal style to improve trust and communication without sacrificing my core values and trying to be who I am not.

For the next several months, Lee had opportunities to watch me in meetings of various sizes, engaging with all levels of management and employees. Afterwards, he was right back to his Socratic coaching style, asking me questions about my intention in one interaction after another: Had I noticed the difference in one person's reaction versus another's? Did I intend to suggest there would be no discussion on some topic – that I already knew the answer? Was I aware that many of the people in the room shut down at that point?

Whether we were reviewing feedback, interview results, Lee's observations from meetings or my description of some interaction he hadn't witnessed, it always followed that pattern. The coach listened, said it back to me from another perspective, and asked if that's what I meant. If not, what *did* I mean? How did I think that I came across? Why did I react the way I did to someone else? Was it consistent with what they actually said, or was it just my interpretation? What could I have done for it to have been more successful – for the interaction to have better matched my intention?

Every discussion ended with Lee summarizing my own answers back to me. Every judgement came from me, not Lee, as he patiently helped me compare my goal to improve to my intentions and my actual behaviours. As he described some contentious exchange he'd witnessed, Lee would often say, 'This is what I hear you saying you mean, but this is how you came across. I know you are sincere in what you say, but most people don't have the benefit of as much conversation with you as I have about your intentions. So they can only react to what they see, which may not be what you intend.'

Lee's coaching helped me realize I could be more *deliberate* – sometimes more *careful* – in how I worked with other people, just as I was in my engineering work. I had already accepted that 75 per cent of the population reacts to *how* I say something and that I have the same tendency with them as well. But I also realized I could make style changes that would limit perceptions interfering with intended communication, if I could remain more deliberately mindful of my own reactions and intentions.

After a few months, our conversations shifted from intentions to results. He cited specific examples that he had heard from my peers or seen for himself. For example, 'I know that you had strong feelings about the topic that was discussed today. Did you notice how different the reaction was than in previous discussions? You appeared more open, and so did the others around the table. Here's what I saw you doing differently . . . Were you aware of it? Were you doing that intentionally?'

The answer to all of those questions was 'Yes!' I had progressed from awareness to deliberately working on style, to aligning the way I came across to better match my intentions – to mindfulness. I could take my time to engage on contentious topics, be more deliberate about tone, expressions, body language etc. I found I could do it without trying to change *who* I was and what I believed to be right and wrong. And I found I could succeed in this improvement; I could increase my mindfulness. As I did, my appreciation grew for a wider range of styles and strengths that do not come naturally to me.

Within six months, Allen Flynt was asked by several senior managers from other organizations, 'What did you to do to Paul Hill? He is so different and such a positive part of the team!'

Lee followed up with another round of senior-manager and customer interviews. Again the responses included some critical remarks, but were overwhelmingly positive and constructive, and they included comments about my strong executive potential. As further evidence for how effective this kind of deliberate mindfulness can be, and how far it can go to changing perceptions, I was promoted to Deputy Director of MOD six months later. Less than a year after that (and only two years after having been persona non grata) I was promoted to Director of Mission Operations when Allen was recruited into private industry.

▸ Personality assessments and feedback can provide great insight and lead to awareness, but improvement must be grounded in behaviour.
▸ Executive coaching can further this awareness and help aspiring leaders learn to be more deliberate in personal style and in aligning behaviours and perceptions with intentions.
▸ As this mindfulness increases, it becomes more natural to moderate style without fear of compromising core values.

Keep Me Honest – Transparency

After I was promoted to Deputy Director, I was invited to speak at a leadership seminar for emerging leaders at the Johnson Space Center. I told them my story of leadership success to persona non grata, and then to gaining awareness and improving my mindfulness in such a short period of time. Afterwards, the instructor, Dr Walt Natemeyer, handed me Marshall Goldsmith's *What Got You Here Won't Get You There* and said, 'I think you'll find the ideas in this book are very relevant to your experience.'

Like the MOD BOD epiphanies during our book clubs, I had another series of 'that's how you say that' moments while reading Goldsmith's book. Just a few of the key insights into gaining personal awareness and mindfulness can be found in these quotes from the book[39]:

- The paradox of success arises from four key beliefs which help us become successful and can be barriers to change:
  1: I Have Succeeded – Successful people believe in their skills and talent.
  2: I Can Succeed – Successful people believe that they have the capability within themselves to make desirable things happen.
  3: I Will Succeed – Successful people have an unflappable optimism.
  4: I Choose to Succeed – Successful people believe that they are doing what they choose to do, because they choose to do it.
- One of the greatest mistakes of successful people is the assumption, 'I am successful. I behave this way. Therefore, I must be successful because I behave this way!'
- The challenge is to make them see that sometimes they are successful *in spite of* this behavior.
- The more we believe that our behavior is a result of our own choices and commitments, the less likely we are to want to change our behavior. It's called cognitive dissonance. The more we are committed to believing that something is true, the less likely we are to believe that its opposite is true, even in the face of clear evidence that shows we are wrong.
- We accept feedback that is consistent with our self-image and reject feedback that is inconsistent.
  1. It is a whole lot easier to see our problems in others than it is to see them in ourselves.
  2. Even though we may be able to deny our problems to ourselves, they may be very obvious to the people who are observing us.

Although I had already embraced this awareness and been promoted into NASA's executive ranks, I was still working to improve my mindfulness. *What Got You Here Won't Get You There* gave me that next push into being able to articulate what I was struggling to apply. Like *Good to Great* and *Built to Last* would soon do for the MOD BOD, Goldsmith's book helped me be even more deliberate and explain it to others, both in Dr Natemeyer's leadership seminar and in the MOD BOD discussions that we had only recently begun. *What Got You Here Won't Get You There* offered a more detailed understanding of the habits successful people fall into that also get in their way, and it also shows how to deliberately change those habits and improve.

The timing could not have been better in my personal evolution and discovering this book. When I made these first significant steps in my personal awareness and mindfulness, MOD management had just embraced John Kotter's *Leading Change* and started to increase the team awareness about our management practices. I was now showing up as the Deputy Director with a fresh appreciation for intentions versus practices in my personal conduct. I took Marshall Goldsmith's advice and told the entire MOD BOD what was on my mind:

> Folks, I know I come across as intense. Some may use stronger words. [pause for their laughter] I know that I have some tendencies that can give you impressions like I'm not going to listen to you or that I think I know everything. It probably won't surprise you to know that, when I take a personality assessment like the Birkman, I turn out to be very mission driven and less people- or relationship-oriented in pretty much every situation. [pause for more laughter] A lot of what you experience with me is related to this, but I'm not making excuses or telling you, 'That's just how I am, so put up with it.'
>
> I know that all of this – what we do, what MOD does – is not about me. It's about us. It's about MOD's responsibility for the cause –

manned spaceflight. Regardless of how I may come across to some folks, I've never had any illusion that I could do any of this on my own, whether it was as a Flight Director or as a manager. Being the boss doesn't make me important – just powerful and dangerous. The truth is, I've always worried that I would fail the team – that I would lead us to fail – and it would be because of me and some mistake I made.

Look, I admit that I'm very passionate about what we do. What can I say? I believe in what we do, that we can make a difference if we lead well, and that we owe it to our people and our predecessors to keep trying to make that difference. For the most part, I believe that kind of passion is important. But that passion and my own weak areas can also reinforce some old and well-earned perceptions. I get that. The passion is part of me, but I really am working on some of the side effects and weak areas.

Trust me when I tell you this. You can help me. Tell me when I'm making your job harder because of how I'm engaging with you. Tell me if I'm not taking the hint and helping you when you need it. Trust that my real focus is doing the right thing, and I know it's you and your people who have to make the tough calls and do the real work.

A year earlier, I couldn't have said those things to the MOD BOD, either as a group or privately. I had not yet come far enough in my own awareness to distinguish intentions from behaviours and style from core values. As I described after Tim Andrews helped me gain my first awareness in this area, I would have been stuck mistaking the goal to improve with the need to accept that I was some kind of bad guy who now needed to be someone else. Just trying to put that into words would have made me angry, and I wouldn't have tried, especially after the successes I had led and the huge efforts they'd required. Instead, as I talked to the MOD BOD on this day, my chief concern was that most of them wouldn't take me at my word; I wouldn't be able to change some long-standing impressions. Although sharing this kind of deeply personal insight felt like standing naked in the room with them, I knew it was the right thing to do – I just had to muster the courage to say it!

Darned if that speech didn't have the same effect that the first six months of my personal evolution had: the MOD BOD took me at my word and helped me! Marshall Goldsmith was right! From time to time, one of them followed me into my office to offer an observation. 'Hey, I know you're working on this, but this is how that came across . . .' It was almost like having a team of executive coaches making the observations Lee Hayward would have made if he'd been with us every day. Our one-on-one meetings took on similar patterns, with much of the emphasis focusing on our relationship, trust, two-way communication, and what I could do to help them.

Once again, it was working. I was still me, but I was becoming a more effective leader for them. They were helping me improve my mindfulness, and it was reflected in the improved 360 feedback I received from all of my direct reports. As their feedback reinforced my improved mindfulness, it also made me more open to the ideas the MOD BOD would soon study in *The Speed of Trust* and work to apply in our management practices.

What Got You Here Won't Get You There had such a profound effect on my personal evolution that, a few months later, I inserted it into the MOD BOD's book-club process, outside of our management-retreat cycle. I joked at the time that I was projecting my sins onto all of them, but they each had a reaction similar to mine when they read it. As we'd discovered in our other book-club discussions, we each had takeaways that were more impactful to us individually, but we all learned something critical that we could apply to ourselves and in our leadership roles.

Some of us posed tougher challenges than others in transitioning from real-time technical leaders into effective managers. After all, despite our similar professional backgrounds, we're different people with different make-ups. We have different blind spots, weak areas and natural tendencies. So it makes sense that helping each of us gain an awareness of them and a willingness to do something about them didn't have a 'one size fits all' answer. However, in the same way the MOD management team first had to gain awareness about our intentions versus our practices, we each had to have some degree of awareness of our individual challenges to help improve our individual mindfulness. That awareness helped us work around our individual weak areas and blind spots. We could then more deliberately match our personal behaviour and communication styles to our intentions.

Personality assessments and executive coaching offer perspectives that many of us would have difficulty seeing for ourselves, and therefore can be an effective first step towards gaining awareness. *What Got You Here Won't Get You There* helps us better understand that this is often an issue of style, not core values. We are then better able to articulate our individual challenges, goals and areas to work on. As many MOD leaders and I learned, we can gain awareness, improve our mindfulness and lead more effectively in a high-trust environment.

- The ideas from *What Got You Here Won't Get You There* offer insights into self-awareness and mindfulness that can help successful people deliberately change less productive habits and become more effective leaders.
- We each have different blind spots, weak areas and natural tendencies; therefore the specific awareness, mindfulness and changes we each need to improve as leaders are different also.
- As we each gain awareness about our individual intentions versus our behaviour, we are able to be more deliberate, increase our mindfulness and lead more effectively in a high-trust environment.

CHAPTER 9

Demonstrating the Morality
as Managers

True discipline requires the independence of mind to reject pressures to conform in ways incompatible with values, performance standards, and long-term aspirations . . . the only legitimate form of discipline is self-discipline, having the inner will to do whatever it takes to create a great outcome, no matter how difficult.

Jim Collins, Great By Choice [40]

Passing Judgement

By the end of the leadership evolution described in Chapter 7, we had come a long way from practices we judged harshly in Chapter 5. The leadership team that did not previously reflect our working-level morality as managers had learned to embrace a very deliberate leadership morality. Although we said it differently in the management ranks (transparency, values alignment and engagement), this was the same alignment to core purpose and high trust (technical truth, integrity and courage) that defined the real-time morality.

The key to making any progress was first finding the willingness to pass judgement on ourselves and our management practices. We gradually came to understand that putting the ripples back in the pond wasn't an act of disloyalty to the organization or each other, but it was absolutely required if we were to manage the problems causing those ripples and lead our team to continued success in this difficult business. With that understanding, critical judgement became less threatening, and the actions required to improve our management practices likewise became easier to identify and implement. We weren't making personal judgements about each other, but we became adamant in holding each other accountable to this transparent, values-driven, fully engaged style of managing. It set the tone for every discussion, whether the decisions entailed our technical work, strategic risk, business management, personnel development, or any subject.

The more we engaged with each other in these terms and saw growing evidence of technical and strategic improvement, the more natural it became, and the less it was seen as judgement or criticism. Rather than giving in to the pressures and nuances of the management cloud, our high-trust environment enabled us to confront our shortcomings and errors, make the necessary judgements, and then take action. We came to see that this was doing our jobs as leaders, and what we owed our customers and stakeholders whose risks we managed: focus exclusively on our core purpose and engage in the most effective way possible to succeed in every decision.

The willingness to pass this kind of judgement in our management practices thus became the enabler in our growth as a leadership team in the same way the real-time morality is the enabler to highly reliable performance in the workforce. This high-trust leadership environment was, therefore, not only the litmus in our decision-making, but it also became the cultural norm and expectation. This went way beyond just me as the boss. Although I expected transparency, values alignment and engagement from all of the senior leaders, they also demanded it from each other and from me. It had taken hold as the overriding morality in the leadership team.

To illustrate how this morality influences a leadership team's effectiveness, consider the set of gears in diagram number 1 on page 229. Three gears are arranged in a gear train, with the largest gear lifting a load. Like winches and similar systems, we crank the smallest gear, and each next larger gear increases the applied torque as it is driven by the smaller gear. This enables the system to lift a much higher load than would be possible with fewer gears.

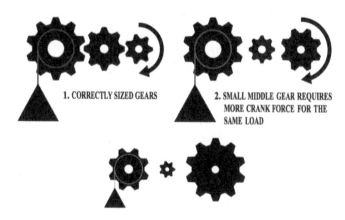

1. CORRECTLY SIZED GEARS

2. SMALL MIDDLE GEAR REQUIRES MORE CRANK FORCE FOR THE SAME LOAD

3. EVEN SMALLER MIDDLE GEAR REDUCES THE LOAD THAT CAN BE LIFTED

Shrinking the size of the middle gear increases the crank force required to turn the gears while lifting the same load, as shown in diagram 2. The system can overcome some change like this, but it takes more effort to make it work.

However, if we shrink the middle gear further, we reach the point shown in diagram 3, in which the crank force required to turn the gears is too high – we can't crank it. In order to turn the gears at all, the load must be reduced. If our goal is to lift as large a load as possible, shrinking the middle gear clearly works against us.

Consider the same system as a model for the real-time morality in the diagrams below. The load we're trying to lift is our core purpose – in Mission Control's case, protecting the astronauts and mission success. The first gear is technical truth, then integrity, and finally courage.

The leadership team dominates the middle gear, integrity. That isn't because only leaders have integrity. Consider who determines acceptable or expected behaviours: the boss or the senior person on any team. In diagram 4, if the leadership team reflects our integrity in all decision-making (pursuing technical truth in every decision and ensuring that the answers align to our core purpose), then we minimize the courage required to make the system work and reach maximum technical truth and mission success.

However, as leaders allow the other 'pressures and barriers' in the management cloud to interfere with management practices, the integrity in their management erodes as the shrunken middle gear in diagram 5 represents. 'No ripples in the pond' and 'diplomacy over clarity' can be overcome, but it requires someone to muster the courage to challenge the boss and the leadership team and overcome the increased resistance in the system. The system can accommodate some of this and still achieve mission success, as long as someone is willing to speak up and face a level of leadership that doesn't want to hear it – to make a difference in spite of the leadership team's practices.

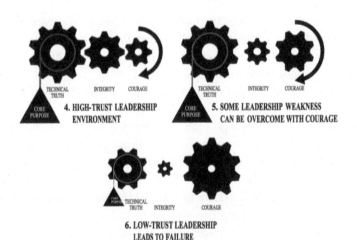

TECHNICAL TRUTH   INTEGRITY   COURAGE

CORE PURPOSE   **4. HIGH-TRUST LEADERSHIP ENVIRONMENT**

TECHNICAL TRUTH   INTEGRITY   COURAGE

CORE PURPOSE   **5. SOME LEADERSHIP WEAKNESS CAN BE OVERCOME WITH COURAGE**

CORE PURPOSE   TECHNICAL TRUTH   INTEGRITY   COURAGE

6. LOW-TRUST LEADERSHIP LEADS TO FAILURE

As the effects of the management cloud further erode our management practices' connection to the real-time morality, the system eventually fails, as shown in diagram 6 on page 230. Expected management behaviours ('reveal as little as possible, don't show your cards, don't throw stones . . .') become strong enough barriers that the courage required to speak up is too high. The culture itself now intimidates someone who could have helped us see our errors and maximize our success. Instead, they choose not to speak up or are shouted down, so we accomplish less and are at great risk of failure.

In contrast, the high-trust leadership environment inherently delivers maximum success by maintaining high integrity and minimizing the courage required to overcome the resistance *from the leadership*. The point isn't for our people to have low courage – it's that our normal way of doing business shouldn't *require* higher courage in order to function consistently with our core purpose and values.

In learning to pass judgement on ourselves and our management practices, Mission Control learned that it was we as leaders who are the keepers of the practices that build or erode integrity from the top down. Focusing on the deliberate integrity in our management practices maximized the trust in the organization. Everyone downline from us at all levels was freed to innovate and accomplish great things, and it became easier for them to challenge our decision-making and ensure that we remained aligned to our core purpose and the real-time morality.

- ▸ The willingness to pass this kind of judgement in our management practices became the enabler in our growth as a leadership team in the same way the real-time morality is the enabler to highly reliable performance in the workforce.
- ▸ A management team that focuses on the deliberate integrity in their management practices maximizes the trust in the organization.
  - ▷ The workforce – everyone downline from you, at all levels – is either freed to innovate and accomplish great things or confronted with mustering the courage to overcome your leadership.
  - ▷ The point isn't for our people to have low courage – it's that our

normal way of doing business shouldn't require higher courage in order to function in a manner consistent with our core purpose and values.

▷ Otherwise, as the integrity in our management practices erodes, we can lead a team into avoidable failure.

Real Results

Although this judgement helped us see the direct connection between our leadership environment and the effectiveness of the organization, we found in Chapter 5 that leaders who profess shared values and good intentions are still at risk of inhibiting highly reliable decision-making in their organization. Thus, while high-trust leadership values – 'all cards on the table' and 'clarity over diplomacy' – became the cultural expectation in our senior-leadership team, it was the resulting *management practices* as behavioural expectations that led to real change and reinforced our alignment as part of our normal business.

The following tables summarize our management practices before and after embracing a high-trust leadership environment. Note that it mimics the table in Chapter 7 summarizing the cultural behaviours of low- and high-trust organizations that Stephen M.R. Covey describes in *The Speed of Trust*. The stark contrast between corresponding practices shows the deliberate evolution to the high-trust style we found through transparency, values alignment and engagement.

Mission Control Management – Basic Governance	
2004	**After the Evolution**
Focus predominantly on safety-of-flight issues, flight readiness and flight control, and control-team technical performance.	Focus on alignment, transparency and engagement in all cost, schedule and technical performance – coordinate across and up the chain.
Technical consistency between divisions was provided almost exclusively during preparation for a specific Shuttle flight or by Flight Directors outside management forums.	Expectation to coordinate all significant changes to existing operating philosophy, flight techniques, procedure or process in ongoing management forums.
Organization-wide panels were primarily limited to flight production/readiness, with little other cross-division collaboration.	All work is managed through organization-wide boards and panels that are delegated to individual divisions or offices. All divisions are expected to coordinate issues through those boards and panels.
The Director's direct reports participated in daily telecons, submitted weekly activity reports and participated in weekly management-info meetings.	The Director's direct reports comprise the MOD Board of Directors (MOD BOD), who participate in all decision-making and strategic efforts.
Technical and resource issues were typically resolved between a division and the directorate in private.	All decisions affecting topics described above in the coordination policy are made with the participation of the full MOD BOD.
Recognition opportunities were dispersed across all divisions and offices, civil servants were given preferential assignments and rewards for top performers were often limited to eventual promotion.	Recognition and awards are prioritized to top performers and role models from the entire organization, and coordinated and vetted by the full MOD BOD.

Mission Control Management – Basic Governance	
2004	**After the Evolution**
Personnel decisions, promotions etc were worked out between individual divisions and the executive office.	Personnel policies, promotions etc are coordinated and vetted with the full MOD BOD.
It was normal to have division-unique, duplicate solutions for common problems.	Management processes are consistent across all divisions, except where work content warrants something unique, and the MOD BOD vets that unique requirement.
A financial baseline existed only at the top executive level, sometimes integrated by the Director personally.	A financial baseline was formally established at the MOD BOD level, shared throughout the management team, and configuration controlled.
No consistent division approach to cost Bases of Estimates.	Fully loaded, division Bases of Estimates are rolled up and traceable to the financial baseline.
Investments were largely 'first to the trough' in private discussion in the directorate office, governed by the available budget – spend what you've got on what you see fit, as defined by each division.	All new investments must have a business case with cost and/or technical ROI ramifications before MOD BOD review and approval.
No baseline configuration control.	Formal change requests are required for baseline changes, and all are reviewed by the MOD BOD.

Mission Control Management – Basic Governance	
2004	**After the Evolution**
Primary resources book kept by the directorate, with margin sometimes parsed out, sometimes used as a 'repository' for a customer. Some secondary resources worked at the division level with no insight by the directorate.	All resources are managed in a custom, directorate rolled-up database that is designed to generate all business reports and financial-plan submits at all levels of management.
A directorate office existed for center business-office interaction with little formal management process and authority.	A directorate office is responsible for business-management processes, reporting and engagement with center business functions.

Mission Control Management – Directorate Level Norms (focused on conflict avoidance)	Mission Control Management – MOD BOD Expectations (focused on transparency, values alignment and engagement)
2004	**After the Evolution**
Best practices were not shared across division lines or with the directorate.	The MOD BOD routinely reviews best practices and failures.
Diplomacy over clarity: Reveal as little as possible of any challenge to the directorate and other divisions; fix your own problems or risk getting 'help'.	*Clarity over diplomacy:* All cards face up on the table for the full leadership team.

Mission Control Management – Directorate Level Norms (focused on conflict avoidance)	Mission Control Management – MOD BOD Expectations (focused on transparency, values alignment and engagement)
2004	After the Evolution
Divisions were concerned primarily with their own area of responsibility and success, not necessarily with the organization's – most divisions saw themselves in competition with the other divisions, rather than working towards common MOD goals.	Divisions are responsible for managing their work and developing their people in a manner consistent with the organization's priorities and the context of a MOD BOD-led effort.
'No ripples in the pond': Divisions were expected to 'do the right thing' but take no actions that could generate concern outside MOD.	1. Divisions and offices are expected and left to operate within the bounds of MOD BOD decisions, directorate priorities and per the rolled-up and allocated MOD-resource baselines. 2. Divisions are expected to alert the MOD BOD about all decisions, positions and issues known or suspected to create concern among our customer executives, external organizations or higher-level executives.
Don't 'throw stones' at other divisions' technical performance and personnel; just manage yours as you see fit.	Be a very tough judge of your individual performance, your organization's and your people's – weigh in on all directorate performance.

As a more concrete example of what these practices brought us, consider our attempts to reduce costs starting in 2004. As we saw in Chapter 5, from 2004 through 2006 we failed in every attempt to rally the management team around the change that was required to make any progress. But our results changed when our high-trust leadership environment developed around us in 2007 and 2008.

It started by applying what we had observed in our bench-marking of the USAF's 50th Space Wing. As mentioned in Chapter 7, the MOD BOD struggled to take real steps towards combining our flight-control and astronaut-training cadres in a move we first called 'Top Gun'. The notion was simple: start everyone in flight operations, develop them into progressively more complex operations, and then develop them into trainers, both for astronauts and flight controllers. Our most skilled experts were those who achieved certification across all operations and training positions in their areas. While that seemed straightforward to some, it was seen by many as too fundamental a change to the way we had always done things.

The remarkable thing about our ultimate ability to make this change is that it did not come by direction from the top down. Through a series of candid MOD BOD discussions, we sifted through the risks and the benefits as a team. In the end, it was one of the division chiefs who made the clear case that it was the right thing to do, both as a better way to develop our people and because it would enable us to downsize while providing the same level of support to our International Space Station customer. Hearing this strategy described so convincingly from one of their peers, the other division chiefs took a new look at their own divisions. Over a few months, they offered similar observations in their divisions: areas that could be combined, or work that was no longer critical, or in some cases, even relevant – we had just 'always done it that way'.

At the end of the year, the MOD BOD committed to a much broader plan that yielded a 23 per cent reduction in the size of our International Space Station workforce as we phased in the changes from 2009 to 2011, as shown on page 238. To our ISS Program customer, that meant a substantial rebate in all of our future costs, with no reduction in the service we provided.

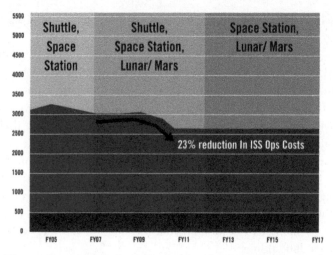

The number of people in MOD, showing the 23% reduction in ISS staffing in 2009[41]

Needless to say, when the ISS Program Manager shared his thoughts with us in our 2009 management retreat, he was very happy about our financial performance. He knew we had been working on a wide range of internal changes, but he didn't expect this kind of help in his own budget challenges. Further, the managers of our other customer programs and some of the executives at NASA Headquarters also took notice that something was changing in MOD, because no one voluntarily gives back excess budget.

We couldn't have pulled it off without having first embraced the high-trust leadership environment, aligning as a leadership team, passing judgement and taking action with the benefit of the real-time morality. From the executive level, we could not have directed the detailed changes required at all levels of the organization to make Top Gun work. The organization and our work were too complex to do that from the top down without great risk of hurting performance in our most critical work. However, a leadership team who no longer saw each other as competing business units was able to guide their divisions through all of the changes while ensuring that we continued to perform at the highly reliable levels our work demanded.

▶ While high-trust leadership values are an essential cultural expectation for a senior-leadership team, it is the resulting management practices that lead to real change and reinforce alignment as part of normal business.

▶ Our leadership team, who no longer saw each other as competing business units, was able to guide their divisions through significant changes while ensuring we continued to perform at the highly reliable levels our work demanded.

Higher Stakes

Our timing couldn't have been better as we hit our stride with these management practices and a very high trust level in 2009, because our commitment was soon put to the test. Almost everything changed in the beginning of 2010, when the Moon and Mars Constellation Program (intended to replace the Shuttle program) was cancelled. With it went one of two major customers, half of our budget and half our workforce after we flew the last Shuttle.

With all of us facing this same slash to our portfolio, the human-spaceflight community across NASA descended into chaos. The prevailing response was to seek any other possible revenue source in an attempt to replace what was lost in our individual budgets. The conversation that played out repeatedly at the Johnson Space Center over the next year and longer was no different, with a focus on keeping as much budget and as many people as we could.

However, the MOD focus was different. We had 18 months of Shuttle mission planning and six final Shuttle flights in front of us. Almost 90 per cent of our Shuttle workforce would be laid off after the last mission, and they knew it. As a result, our first concern was to ensure we flew every one of the remaining Shuttle flights with the same level of performance we always had, recognizing that the closer we got to the end, the greater the risk that our people would lose focus – if not leave altogether. After that, we worried about what else we risked losing if MOD no longer conducted the high-paced launch-and-entry operations. Rather than trying to do whatever odd jobs we could find, MOD focused on 'core values and enduring purpose', as we learned from *Good to Great*. We strategized on the work we needed to capture to ensure that we maintained the technical skills and culture necessary to fly outside of low Earth orbit if so required in future.

We had our work cut out for us. As mentioned in Chapter 5, the ISS Program Manager who had just been singing our praises for reducing costs by 23 per cent was now criticizing us as part of the ongoing NASA problem. He wasn't alone in that criticism. Many of the executives at NASA Headquarters shared his opinion. A significant reason the Moon and Mars program was cancelled was to shift more of the responsibility away from NASA organizations like ours in the hope of reducing costs and breaking out of doing things the way we always had.

In the short term, we embraced two specific points from *Good to Great*[42]:

- Yes, leadership is about vision. But leadership is equally about creating a climate where the truth is heard and the brutal facts confronted.
- Stockdale Paradox: Retain faith that you will prevail in the end, regardless of the difficulties, *AND at the same time* confront the most brutal facts of your current reality, whatever they might be.

This situation was bad. We knew it, and we said so. If we didn't respond effectively, we could lose focus, screw up and let our errors take the lives of an upcoming crew of astronauts who were relying on us. If we did not change some parts of the NASA strategy, we were going to stop performing some of the more difficult phases of flight, and we would lose the experience and skills that had been honed for decades. Like we had learned from our benchmarking, it was now up to us to do something about these risks, or it would be because of us (if we did nothing) that we lost out.

We didn't reserve this candour for private MOD BOD discussions. We talked in these terms to the entire MOD workforce in large 'all hands' meetings. The goal was to acknowledge to them that the leadership got it: we saw the threats, and were fully engaged in ensuring MOD did not lower the bar in support of our astronauts while also trying to influence the longer-term outcome. During this candid discussion with the workforce, we also made it clear that, even in a best case, we faced laying off more than half our Shuttle workforce.

The MOD BOD then set off on a 3-part, long-term strategy:

1. Take the next steps to continue reducing our costs.
2. Reach out to our stakeholders to ensure that they understand our management accomplishments, not just our rocket science achievement.
3. Seek work that fits and could be critical in preserving our core capability.

How did we do?
1. By the end of the year, we had developed a project to re-engineer the vast computing and communications systems in the Mission Control Center and spacecraft-simulation complex. When our new systems went online, they cut our fixed costs in half, which reduced our total cost of doing business by another 25 per cent. We also took more steps to reduce the size of the workforce required to conduct operations for launch-and-entry vehicles.

2. In discussions with our customers, executives at NASA Headquarters and industry executives, we successfully demonstrated our management changes and the dramatic cost reductions we had accomplished while delivering more complex work. As a result, MOD won back the operations responsibility for the new spacecraft and rocket NASA continued developing (Orion and the Space Launch System, SLS). Counter to the original intent of NASA's strategy, MOD was also allowed to pursue work from the companies from whom NASA intended to buy commercial launch services.

3. MOD negotiated with three companies to provide operations services to their commercial space-launch services. Of those, Boeing's CST-100 spacecraft was chosen as a commercial service to fly NASA astronauts to and from the International Space Station. Boeing selected none other than NASA's MOD to be their flight operations team, working like a subcontractor to Boeing.

The combination of all of these yielded the budget profile shown in the figure below. By 2015, the changes we made in pursuit of our strategy reduced the size of MOD's workforce to half the size we had already come down to by 2009. In fact, MOD had contracted to a fourth of the size we had been in 1990 while flying Shuttle and just beginning preparations to build ISS. The savings were permanent rebates to our customers for the same level of work. But at this level, we weren't going out of business; as summarized above, we were flying ISS, developing flight test operations for Orion and SLS, and developing operations as part of Boeing's commercial CST-100 team.

Along the way, we kept the workforce informed of our progress in capturing critical new work. As predicted, even with the new work we had won, we were still forced to lay off 80 per cent of our experienced Shuttle workforce. The workforce trusted us to fight the good fight and matched us in kind by staying until the end of the last Shuttle mission, in July 2011. The vast majority of those we couldn't keep on board cheered us as they went out the door and thanked us for keeping MOD in the game, even if they could no longer be part of it.

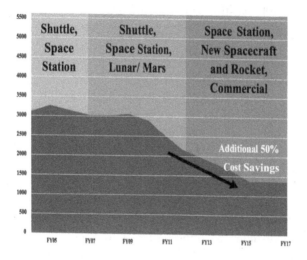

The number of people in MOD, showing another 50% reduction, 2011-15[43]

Many people mistook these dramatic cost savings as our primary goal and accomplishment, but our sights were set much higher. It is true that our cost performance and openness were key to changing our stakeholders' attitudes towards MOD; where they had previously respected only our technical capability they now trusted our business management. As that trust increased they became more and more willing to consider MOD as part of the overall solution going forward, not just part of what they saw as an obsolete and inward-looking past.

Thus, our real coup was more fundamental than cost reductions; it was winning back the launch-and-entry services for the new NASA rocket and spacecraft, and being permitted to provide our full range of services as a Boeing commercial-operations team member. This meant we had secured the exact work that was critical in preserving our core capability for the day when the nation asks NASA to return to the Moon, go to Mars, or anywhere else outside of low Earth orbit. We had secured the work that would reinforce our ability to deliver on our core purpose of protecting astronauts and accomplishing new missions.

If we had faced this challenge without having evolved our leadership environment, we would have simply circled the wagons and tried to protect what we had: budget levels, facilities and equipment, and the number of people in the organization. Like other affected organizations after the 2010 cancellation, we would have grasped at straws for any work possible just to keep our roster and dollar value as high as possible.

We would have failed. There wasn't any money in the federal budget for us to keep more people on the payroll without a mission-related program that required our support. Our International Space Station customer didn't have the resources to essentially pay us double for the same amount of work. If we had tried this route, we would have only confirmed our customers' and stakeholders' criticisms that we were part of the problem and not able or willing to innovate and control costs.

Instead, this highly transparent, aligned and engaged management style had become so habitual the MOD BOD focused on preserving our mission focus and the culture we consciously attributed with our ongoing success. It was almost second nature, and not just to me but to the entire MOD BOD.

- ▶ Under extreme pressure, and with the Stockdale Paradox reminding us to confront the brutal facts, we didn't just weather the storm; we implemented innovation after innovation and strengthened our deliberate leadership culture.
- ▶ Mission Control's real coup in stewarding the real-time morality in our leadership team was more fundamental than cost reductions; it enabled us to secure the work that would reinforce our ability to

deliver on our core purpose of protecting astronauts and accomplishing new missions.

Stewardship

The specific steps we took in making technical improvements and huge cost reductions originated within the divisions, not in the senior-leadership team. I didn't dream up the specific technical changes and direct them from the top down. At most, I was a catalyst for the MOD BOD to each keep a top-level, MOD perspective, and I listened for signs that any of us was falling back into the old, stove-piped way of doing business. As the MOD BOD took that same focus down into their divisions, their people brought them one great idea after another. The MOD BOD then simply assessed the cost and risk of each change and decided whether we were ready to commit. Each success led directly to, and reinforced, the next.

The fact that developing and executing our strategy came naturally doesn't mean it was easy. Although we were reducing costs and increasing capability, decision after decision meant widespread budget cuts and reducing the size of the workforce. We were making choices that, even if successful, would require us to lay off significant numbers of our people who looked to us to protect them. Each one of us knew the price we were paying and the impacts it would have on the workforce. We were going to leave some part of our MOD family behind as we took each next step, and we all carried a heavy emotional burden. It was not uncommon for one or more of us to balk at some specific change. Perhaps some change was a bridge too far on that day, or the next 5- or 10-person reduction in some leader's division just seemed too disloyal to the people who had served us so well.

However, staying the course – keeping the focus on our core purpose and culture – was almost always a self-policing event in the MOD BOD, and just as often by someone else around the table as it was from me. The MOD BOD often reminded each other what was really at stake, and that our priority was to preserve the real strength of MOD, not just another wedge of money or a few people in some specific corner of a division. In almost every case, obstinate MOD BOD members came around, either during the meeting or shortly thereafter. When they didn't, it rarely required more from me than to point out in private that they were losing a MOD BOD focus and sinking back into 'protecting what they had' rather than helping us do the right thing for MOD and human spaceflight.

Similarly, disagreements rarely led to confrontation or controversy at the MOD BOD level. The MOD BOD became adept at anticipating concerns from their peers, and would often already be addressing them before a full MOD BOD discussion. Again, if we ran into some impasse, my role typically was simply to help them get past a roadblock in the discussion and back to a decision based on our shared values and purpose.

Sometimes the MOD BOD had to keep me on course as well. For example, one of our prime contractors developed a prototype ISS simulation that ran on a single personal computer instead of the complex, multi-million-dollar, server-driven systems we used to train flight controllers. I instantly thought about ways we could use this simpler and cheaper simulation to offer more hands-on training to our least experienced flight controllers at a fraction of our normal cost. To the apparent surprise of the engineers who were demonstrating the system, I asked our training team to set up several of these prototypes in our simulator complex for the flight controllers and instructors to evaluate as training tools. They did what they were told, and many flight controllers and instructors spent precious hours testing the prototypes. I was convinced these folks would figure out how to move us to these cheaper devices if they had enough time with them.

Six months later, as we were reviewing progress and costs for this little scavenger hunt, one of the MOD BOD members spoke up and told me, 'Paul, we are convinced this system is never going to be able to do what you have in mind. It's the right idea, but the processors still can't handle the computing load. We're throwing good money after bad to keep pursuing it, and I have more important work for my people to do than working on something we now know is a dead end.'

Just a few years earlier, they would have kept pursuing the idea at just about any cost rather than tell the boss he was wrong. Instead, they reminded me that we had better things to do, and should get on with them. Done. We shut down the work immediately and kept looking for innovations that could actually help. (I was also much more careful in the future when I had a 'great idea'. Rather than directing the team to pursue it, I'd engage the MOD BOD in the dialogue.)

Having watched as the MOD BOD engaged under our new rules, the MOD BOD members' deputies took note. In my periodic skip-level meetings with them, the deputies made it clear that they saw what the MOD BOD evolution had done for us as a leadership team, and what it was doing for MOD. They wanted in: in on the leadership development and in on the cultural discussion. We happily obliged, and then we pushed the leadership evolution down to the next level of management within each division. Our goal for them was the same as it was for the MOD BOD's evolution: join us in stewarding the culture and leading based on core purpose and values. (The process we used to do that will be described in Chapter 10.)

We had come a long way from 'no ripples in the pond'. We brought the real-time morality back into our management ranks, and it strengthened our performance as a management team. We then evolved from deliberate management practices to a real understanding and stewardship of the underlying values that reinforced our strong performance. That stewardship served us well when we were under severe pressure, keeping us aligned to core purpose and values. Ultimately, it brought us success in capturing critical work that we otherwise could not have won.

- An aligned and high-trust leadership team is better able to guide their teams through significant change while ensuring they continue to perform at a highly reliable level.
  - A highly transparent, aligned and engaged management style can become habitual, leading a leadership team to focus on core purpose and the culture that is critical for continued success.
  - Each successful change leads directly to, and reinforces, the next.
  - The fact that developing and executing any strategy comes naturally doesn't mean it is easy, but the focus on core purpose and culture keeps us aligned.
- Pushing the leadership evolution down through each level of management achieves the same goal as the senior leadership team's evolution: each level joins us in stewarding the culture and leading based on core purpose and values.
- Stewardship serves a leadership team well when we are under severe pressure, keeping us aligned to core purpose and values, and enabling critical success.

CHAPTER 10

A Roadmap to Transformation

The number-one job of any leader is to inspire trust. It's to release the creativity and capacity of individuals to give their best and to create a high-trust environment in which they can effectively work with others.

Stephen M.R. Covey, The Speed of Trust[44]

Embrace the Morality

The critical awareness that Mission Control lost and rediscovered in our management roles is that everything we do contributes to the organization's success or failure, not just in our individual jobs but in the organization's core purpose. In the Mission Control Room, it is that awareness that leads to a shared ownership, not just in the outcome but in *how* we perform and *why* we make every decision. We hold ourselves and each other accountable for applying the real-time morality: for asking the 'whys', focusing on finding the right answer – not just winning the debate – and having the courage to speak up and take actions that will best protect the astronauts and accomplish the mission. Passing judgement in this way becomes second nature in the Mission Control Room, which further increases trust and strengthens team performance, especially in intimidating situations, where mistakes can have immediately catastrophic results.

However, as we move away from the real-time decision-making and the risk of immediate catastrophe, the management cloud gradually lures us into also leaving behind that specific, enabling morality. As the responsibilities and pressures in the management cloud increasingly dominate managers' daily experience, they can dilute the willingness to pass the same deliberate judgement that becomes second nature in the Mission Control Room. As a result, *in our management roles*, we risk losing alignment and leaving behind the critical, high-trust behaviours as we make decisions that can cost us customers, bankrupt the organization, impact workforce performance and risk us failing in our core purpose.

Finding our way back from 'no ripples in the pond' to the real-time morality in our management roles first required us to relearn this awareness. Thus, Mission Control's most important lesson from benchmarking wasn't any particular practice we found. It was the realization that we were *contributing* to our problems – not *solving* them. That realization was enough to help us overcome our individual discomfort with passing judgement against our organization, management practices and individual behaviours. Our core purpose and responsibility were still too important to each of us to risk our organization's performance by refusing to change our low-trust management practices.

We clung to that initial awareness as we explored new ideas in our book-club discussions and applied them to our management practices. As our management team evolved, we became more comfortable judging our practices and ourselves, and talking in terms of core purpose and values. It became second nature to pass judgement in management forums in terms of our transparency, values alignment and engagement. As we turned to effective, high-trust leadership and common purpose, the necessary behaviour changes become more obvious, and more natural. Each step we took then reinforced the awareness and our natural tendencies for higher-trust behaviours, just as each element of the real-time morality reinforces the next.

Evolving a team to high-trust leadership hinges on embracing the real-time morality, not just at the working level but also as managers. That is exactly what we accomplish as we learn to remain deliberately mindful of our shared core purpose, transparency, values alignment and engagement, and continuously reinforce that awareness and mindfulness in our behaviour. The management team's culture shifts deliberately to clarity over diplomacy – one that preserves the integrity in ongoing practices – and an environment that prizes continuous judgement based on the shared morality. Rather than allowing the management cloud to dilute the connection to their core purpose, the real-time morality empowers a leadership team to preserve the connection, just as it does for the team in the Mission Control Room.

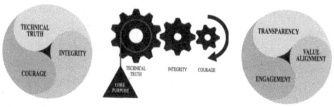

Trust elements of the real-time morality
At the working level (left), in a high-trust environment (middle)
and in management (right)

Evolve Your Team

Embracing the deliberate morality and evolving to a high-trust leadership culture and management practices is repeatable in any leadership team, not just in Mission Control. The morality and awareness empower the leadership team to have the uncomfortable discussions, make critical judgements and choose to be as deliberate in all facets of management as in any critical decision-making. With this focus, the team can then explore ideas to increase trust, understand and articulate core values and purpose, and align as a leadership team. The resulting leadership culture entails high-trust and fully transparent engagement among the senior leaders in the same way that successful spaceflight requires high trust and full transparency from the flight control team.

After the Mission Control management team had made this journey together, our management practices had fundamentally changed to reflect the deliberate cultural shift, as summarized in the Mission Control Management tables in Chapter 9. These practices were the result of both embracing the real-time morality in our management ranks and implementing the specific behaviours that ensured we truly walked the talk – and didn't simply profess alignment.

By leveraging our experience, the journey can be made more deliberately and in less time than we took in our discovery. Consider what you've already learned about our journey as a head start in the awareness that took us years of failed management retreats and benchmarking to achieve. The 4-stage team guide outlined in the next several tables provides the detailed roadmap for stepping methodically through these evolutions, both philosophically and in practice.

The four philosophical evolutions to lead a team through are:

1. Align the team to core values.
2. Incorporate transparency and high-trust behaviours as team norms.
3. Anchor all management strategies and decisions on deliberations in this transparent, aligned and engaged team.
4. Reinforce an ongoing integrity focused on deliberate alignment and high-trust behaviours.

This process relies on the book-club experiences and foundational ideas that were key to Mission Control's evolution, and which are laid out exquisitely in the following five books:

1. *What Got You Here Won't Get You There: How Successful People Become Even More Successful*, Marshall Goldsmith.

2. *The Speed of Trust: The One Thing That Changes
   Everything*, Stephen M. R. Covey.
3. *Leading Change*, John P. Kotter.
4. *Good to Great*, Jim Collins.
5. *Built to Last: Successful Habits of Visionary Companies*,
   Jim Collins and Jerry I. Porras.

The team steps through the series of 'book club' discussions in open, round-table team dialogue. The goal is to explore each book's ideas, their applicability to your workplace, and the team's effectiveness in demonstrating them in actual management practices. These discussions reinforce the ideas and high-trust style in individual leaders, highlight practices that could be changed to further improve the leadership environment, and enhance organizational performance. The evolution leads the team towards high-trust behaviours and a willingness to critically evaluate current practices; towards exploring more ideas and developing an openness to change aligned with values and a common core purpose.

Changing trust behaviours and cultural norms is such a personal experience I recommend going through this one level of management at a time, starting with the team you lead. As you progress through the stages, trust will build between you and the team, and among the team as peers. After achieving a transparent and engaged team as the new normal, members of the team will then be better equipped to take their direct reports through the same evolution as a team, thus pushing the high-trust leadership down each level of management.

The key is open and thorough dialogue in each book-club discussion to internalize the ideas and observations they help us make, followed by deliberate and related management-practice changes and actions. As my executive coach, Lee Hayward, reminded me:

Reading books does not 'make the person' – the journey you took as a person/leader prepared you to be able to receive the concepts in the books. There are so many persons who go through leadership programs, who read the books, take part in the discussions, etc – but these do not make them leaders. The question is, 'Did they get it?' Have they gotten beyond understanding to actually being able to *be* effective? Reading all that stuff informs them but doesn't necessarily make them into the kind of leader you are talking about.

It will then take some weeks or months for the new behaviours to become more habitual, to become the new normal. How long the full process takes will vary. A team that is already amenable to a high-trust leadership environment can progress through each stage in two to three months. Some teams may need more time to become accustomed to passing judgement on their own organization and practices, stretching some stages into 6-month exercises as new practices are adjusted to have the desired effect from each book club. As the evolution flows down the organization, the high-trust leadership environment becomes dominant in the management cloud, and its role models develop at every level.

Stage 1: First Steps Towards Awareness and Alignment

Follow each step in sequence below, starting with Stage 1, which sets the scene for the progressively more engaged and transparent stages that follow. Before debating new goals, core values and management practices, Stage 1 sparks the intentional awareness in this deliberate morality – one that is focused first on transparency and engagement.

Stage 1: First Steps Towards Awareness and Alignment
1. Align the senior leaders (direct reports)
Top Executive reads *What Got You Here Won't Get You There*
Top Executive outbriefs all direct reports on personal takeaways and team observations from *What Got You Here Won't Get You There*

Facilitated retreats – as required in these discussions
Establish the Senior Leadership Team (SLT) of *all* direct reports
Senior Leadership Team review of *The Speed of Trust*
2. Maximize transparency and high trust among the Senior Leadership Team
SLT privy to all issues, philosophy changes, resource changes, non-standard decisions
No more 'decisions' in 1-on-1 meetings with direct reports

Since authority and power resides with the boss, as a first step in aligning the team, the senior leader reads *What Got You Here Won't Get You There* and summarizes their takeaways to the rest of the team. Like the book-club discussions that will follow, the goal is for the boss to point out lessons learned about themselves, how they engage with the team, and their goal to evolve towards a transparent, aligned and engaged team. In addition to the specific observations, the discussion helps set the expectation that this kind of introspection is helpful, and that the judgement that it leads to is acceptable.

If you're not comfortable leading your team through this and the subsequent discussions, bring in a facilitator. It may take time for this kind of discussion and these behaviours to feel comfortable, and any pre-existing trust problems in the team make this tougher still. However, even with a facilitator's help, the goals in each step must still be yours. Along the way, if you're still struggling with discomfort, also consider a personality assessment and some executive coaching to ensure that your engagement with your team accurately reflects your intentions.

Establish your Senior Leadership Team (SLT, *your* MOD BOD), which is all of your direct reports. Their jobs don't change, but the expectations may. Now there is a deliberate goal for full transparency and engagement in all decision-making in your team, and those are expectations for which you will be held accountable as well.

As a first team effort, conduct a book club of *The Speed of Trust* with this team. In an informal group setting, compare notes on the ideas that resonate with each team member, which ideas the team demonstrates, and which they do not. From this, also discuss the existing behaviours the team considers top priorities to change in order to increase trust.

In the last two steps of this stage, we make our first real management-practice changes towards transparency and engagement – putting the ripples back in the pond. Increase the openness and trust by coordinating and ratifying a formal policy that the Senior Leadership Team will share all issues, philosophy changes, resource changes and non-standard decisions from within their areas of responsibility. By committing to 'no one-on-one decisions', you are simply reminding them that this policy applies to you as well.

Stage 2: Deliberate Transparency and Alignment

As the new expectations and behaviours from Stage 1 become more comfortable, the team is ready for Stage 2 – to further solidify transparency, alignment and engagement as the new normal.

Stage 2: Deliberate Transparency and Alignment
1. Align the senior leaders (direct reports)
Deliberate vision and strategy communication
2. Maximize transparency and high trust among the Senior Leadership Team
Senior Leadership Team review of *Leading Change*
Regular, SLT-only meetings: leadership, culture, alignment-related dialogue
Push authority down; bring SLT in on all decision-making

A top-level vision and strategy helps gel the ideas from the Stage 1 discussions into a concrete understanding of the organization's core values, and the underlying intent in evolving the management practices. The open discussions involved in coordinating this vision across the team are as important as the specific results, because it is through this give-and-take that the Senior Leadership Team reinforces engagement and ownership of the strategy itself.

The book-club discussion of *Leading Change* then offers the framework for the team to take on more fundamental challenges and change in the organization. Conducting Senior Leadership Team meetings focusing on the ideas from Stages 1 and 2 leads to progress in addressing the organization's challenges, and further reinforces the team's willingness and effectiveness in judging the team's alignment and performance.

Formally declaring the intent to include the team in all decision-making reinforces the reality of their expanded role. As this becomes more habitual, transparency, alignment and engagement become not just the new normal but an expectation from the team. The Senior Leadership Team will then become progressively more natural in performing as a high-trust team reflecting the real-time morality in their management setting.

Stage 3: Accelerating Team Evolution to Stewardship

Your more naturally transparent and aligned leadership team is now ready, in Stage 3, to turn their attention to core values and the next evolution in deliberate management that both reflects and reinforces values and purpose.

Stage 3: Accelerating Team Evolution to Stewardship
1. Align the senior leaders (direct reports)
Senior Leadership Team review of *Good to Great*
Senior Leadership Team develops priorities and corresponding business strategy

2. Maximize transparency and high trust among the Senior Leadership Team
Succession management requires real SLT collaboration on all leadership-team candidates
Periodic 'surgical' SLT rotations to catalyse values-driven leadership individually
Senior Leadership Team review of *What Got You Here Won't Get You There*
3. Deliberate execution based on integrated SLT strategy
SLT baselines a comprehensive budget and methodology
Integrated organization-budget configuration controlled from top down
SLT reviews cost/schedule/technical performance vs all fund sources

The *Good to Great* book club sets the stage for the team to develop a formal set of organizational priorities based on core purpose and values. In the next step, the team will assess and evolve the business strategies in alignment with the top-down priorities – another concrete step towards ensuring your management behaviours and practices are consistent with your core purpose and values.

In a first step towards making these evolutions permanent cultural changes, add an expectation that the Senior Leadership Team now collaborates on candidates for promotion into any position that reports directly to any member of the team. At least annually, engage the full team in assessing all of their direct reports in current performance and their potential to evolve to high-trust leadership. Simply having the open discussions across the leadership team that focus on the practices that reinforce transparency, alignment and engagement is as important as the individual assessments. Through these discussions, the Senior Leadership Team continues to normalize these high-trust and aligned behaviours.

The book-club discussion of *What Got You Here Won't Get You There* provides the senior leaders with another dose of personal awareness. With the ideas from previous steps already internalized, this furthers their understanding and commitment to management practices that reinforce the high-trust style that embodies the real-time morality.

If your organization hasn't already done so, Stage 3 takes the next steps in formalizing organization-wide, top-down budget-and financial-management processes. Adding configuration control helps the team reduce ad hoc resource changes that may inadvertently work against the team's declared strategy. That awareness, and the periodic cost/schedule/technical review of all work and all fund sources, ensures that the leadership team is deliberately managing any areas of concern, and literally getting the most bang for their buck. They also formally reinforce transparency as a management practice.

Stage 4: Evolving Leadership Values into a Culture

This last stage ensures that the senior leaders are focused on long-term performance and furthering the philosophical focus as the normal culture in the management ranks. As Jim Collins says, this is the next step in the shift from 'time telling' to 'clock building' as the cultural norm. Stage 4 rounds out the team's journey into the culture of the real-time morality.

Stage 4: Evolving Leadership Values into a Culture
3. Deliberate execution based on integrated SLT strategy
SLT insight into baseline and experience-based, standard-support BOEs and proposals
4. Deliberate, recurring 'realignment'
Senior Leadership Team review of *Built to Last*
Self-facilitated annual leadership retreats on priorities, challenges, leadership culture
Select leaders for predisposition to values, behaviours – clock building

Rotate rising middle-level leaders deliberately to facilitate a leadership perspective
Ruthless continuous improvement
No room for the unwilling

The Senior Leadership Team takes one more step in business management by reviewing the basis of estimates and the methodology for each across all business units. As in the previous management-practice changes, the focus in this step is transparency and ensuring the methodologies are deliberately aligned to the organization's core purpose and strategy.

The remaining steps deliberately reinforce the team's values in the leaders they are developing to replace them. This begins with the *Built to Last* book club. By this milestone, the Senior Leadership Team should be able to facilitate its own retreats and those that include the lower levels of management. Ideas that may have once seemed foreign and uncomfortable to address will have become a more normal part of the vernacular and management approach; thus, on many or most subjects, this team can navigate the tough discussions effectively.

The next steps then up the ante on management and leader selection by going beyond collaboration to a deliberate emphasis on the ability to manage with transparency, alignment and engagement. To reinforce this focus before a final selection is made, require the selecting official to review how the top candidates compared in the high-trust behaviours, and why they are seen as the best candidate for a position that puts them one step closer to the Senior Leadership Team.

Rotating managers to equivalent positions in another work area with a different boss and peers helps them focus on leading their team. This is especially helpful for managers who have been promoted up the ranks in the same area in which they excelled at the working level, and who now may struggle to let go of being the working-level expert rather than a values-based leader.

Encouraging ruthless continuous improvement means regularly reinforcing that nothing is sacred except your high-trust leadership environment aligned to your core purpose. Every practice, project or product line is subject to ongoing assessment with the goal to keep challenging the status quo, reduce costs and innovate.

As the last step, it should now be clear to the full team that high-trust leadership requires the entire team's buy-in. It does not require everyone to agree on all specific decisions; however, any holdouts to full transparency, alignment and engagement will reduce trust throughout the team. Encourage open debate on disagreements, but do not tolerate holdouts in the trust behaviours because they will degrade organizational performance over time.

The table below integrates all four stages into a single guide. It is the full journey that brought the real-time morality back to Mission Control management and enabled us to evolve into a transparent, aligned and engaged leadership team. The steps in each stage are indicated by the shaded boxes under each stage number in the right-most columns, and they are performed in the same order as they are when using the previous separate tables.

Team Guide: Evolving Mission Control Leadership Values into a Culture Stage				
Management Practice Evolution Steps	1	2	3	4
1. Align the senior leaders (direct reports)				
Top Executive reads *What Got You Here Won't Get You There*	▓			
Top Executive outbriefs all direct reports on personal takeaways and team observations from *What Got You Here Won't Get You There*	▓			

Team Guide: Evolving Mission Control Leadership Values into a Culture Stage				
Management Practice Evolution Steps	**1**	**2**	**3**	**4**
Facilitated retreats – as required in these discussions	■			
Establish the Senior Leadership Team (SLT) of *all* direct reports	■			
Senior Leadership Team review of *The Speed of Trust*	■			
Deliberate vision and strategy communication		■		
Senior Leadership Team review of *Good to Great*			■	
Senior Leadership Team develops priorities and corresponding business strategy			■	
2. Maximize transparency and high trust among the Senior Leadership Team				
SLT privy to all issues, philosophy changes, resource changes, non-standard decisions	■			
No more 'decisions' in 1-on-1 meetings with direct reports	■			
Senior Leadership Team review of *Leading Change*		■		
Regular, SLT-only meetings: leadership, culture, alignment-related dialogue		■		
Push authority down, bring SLT in on all decision-making		■	–	
Succession management requires real SLT collaboration on all leadership-team candidates			■	
Periodic 'surgical' SLT rotations to catalyse values-driven leadership individually			■	

Team Guide: Evolving Mission Control Leadership Values into a Culture Stage				
Management Practice Evolution Steps	1	2	3	4
Senior Leadership Team review of *What Got You Here Won't Get You There*				
3. Deliberate execution based on integrated SLT strategy				
Senior Leadership Team baselines comprehensive budget and methodology				
Integrated organization budget configuration controlled from top down				
SLT reviews cost/schedule/technical performance vs all fund sources				
SLT insight into baseline and experienced-based, standards-supported BOEs and proposals				
4. Deliberate, recurring 'realignment'				
Senior Leadership Team review of *Built to Last*				
Self-facilitated annual leadership retreats on priorities, challenges, leadership culture				
Select leaders for predisposition to values, behaviours – clock building				
Rotate rising middle-level leaders deliberately to facilitate a leadership perspective				
Ruthless continuous improvement				
No room for the unwilling				

Develop Your Emerging Leaders

We each have the opportunity to increase our own self-awareness and mindfulness, and then deliberately improve our high-trust leadership style. It isn't rocket science, although focusing only on our rocket science can get in our way! Helping our emerging leaders to evolve past that limited perspective creates better leaders and leadership teams who can then lead their organizations to pull off what they once thought was impossible. Developing your strong performers and less experienced managers through a journey like the leadership team's is the next step in deliberate recurring realignment.

The first priority should be to develop the managers who are at the level from which you will promote into your Senior Leadership Team. This primes the pump with senior-leadership candidates who have already evolved their individual-leadership styles and are ready to join and further the Senior Leadership Team's efforts.

In this individual-leadership development guide, we apply the book-club process that emphasizes peer and supervisor discussion of the key points to leverage the same ideas that are invaluable in transforming the senior-leadership team. As the first stage in the emerging-leader development guide says, this starts with selecting people into this level of management who have already demonstrated some awareness and willingness to engage in the transparent style. Since they will be candidates for senior leadership after this selection, it is an important step for your direct reports to compare notes on their observations before any selection. This helps the selecting official be more deliberate in their evaluation, adds peer perspectives they may not have, and reinforces trust within the Senior Leadership Team.

Like the leadership-team evolution, this process isn't rocket science. As managers are promoted into the level that reports to the Senior Leadership Team, give them each a copy of the five books. Schedule book-club discussions with the new manager, their peers and their senior leader (the selecting official). Focus on key takeaways, personal strengths and weaknesses, and areas for both the new manager and the team to work on going forward. Then move to the next step.

In Stage 1, the goal is to deliberately raise the new manager's awareness about their personal style, the effects of the management cloud, and the high-trust environment they are expected to evolve into. As those ideas become more comfortable, Stage 2 raises their perspective to core purpose and values rather than just their expertise or the previous successes that originally got them noticed. After this stage, our more aware, high-trust and values-focused manager is better equipped to lead meaningful and effective change, and Stage 3 formalizes that readiness with John Kotter's help.

Individual Guide: Developing the Emerging Leader into Mission Control Values
Management Practice Evolution Steps
0. Select the right people into the emerging-leadership ranks
Seek top performers, with strong domain competence, who show a predisposition to the values and behaviours necessary to further the evolution in cultural stewardship
Compare observations and perceptions about candidates with peer leaders during selection
1. Raise their personal awareness and leadership perspective, and apply the awareness to catalyse deliberate, high-trust leadership behaviours by having them:
Read *What Got You Here Won't Get You There* and compare notes on the takeaways with you and their peers as a group
Take the Birkman Assessment, and review the results with a trained specialist

Read *The Speed of Trust* and compare notes on the takeaways with you and their peers as a group

2. Increase their awareness of core purpose and values, conscious alignment and focus on cultural stewardship by having them:

Read *Good to Great* and compare notes on the takeaways with you and their peers as a group. Leverage this conversation to discuss the organization's core purpose and values with you and their peers as a group

Read *Built to Last* and compare notes on the takeaways with you and their peers as a group

3. Sharpen their critical eye and their ability to lead change by having them:

Informally assess alignment to core purpose and values, and high-trust behaviours, and identify opportunities for improvement – theirs, in their team, and in the level above

Read *Leading Change* and compare notes on the takeaways with you and their peers as a group

4. Continuous feedback and growth

Regular, informal tag-ups to discuss progress, challenges and help/change needed; focus on their team's role, trust environment and culture as part of the larger organization

Include specific high-trust practices in performance planning/ratings

Selective, lateral rotations to expand perspective and further evolve leadership potential

Selective use of executive coaching for high-potentials, targeting specific leadership growth

5. Deliberately promote them to higher levels of leadership

Full Senior Leadership Team evaluation and ranking of the emerging leaders as a group, emphasizing performance and potential

The final two steps are not separate stages but ongoing practices intended to continue evolving each emerging leader, as well as their more experienced and seasoned predecessors. The goal is to further normalize:

- Regular discussions focused on transparency, values alignment and engagement.
- The willingness to judge individual and team behaviours; and
- Stewardship of these leadership values on behalf of the organization's core purpose.

Including specific practices (for example, coordinating all issues, philosophy change, resource change and non-standard decisions with the full team) in written performance plans as part of each leader's formal evaluation holds you and your direct reports accountable to the behaviours, and to regular feedback. Lateral rotations serve the same purpose as they do in the team guide: help the emerging leader let go of being the working-level expert and instead become a values-based leader. For high-potential emerging leaders who struggle with some parts of this evolution, executive coaching may help them understand their weak areas and blind spots.

Evaluating the emerging leader on an annual basis with your full senior-leadership team in a format similar to the book-club discussions fosters a focus on high-trust leadership potential as this leader is considered for promotion to the senior level. This helps each supervisor articulate strengths and areas for improvement for each emerging leader. It also raises the awareness of senior-leadership team members of the leadership potential of a broader pool of candidates. The periodic group evaluations lead naturally to selecting candidates who are most prepared for high-trust leadership. This focus reinforces culture stewardship based on core purpose and values.

Incorporating Steps 4 and 5 as regular practices reinforces the management practices evolved through the team guide. It also reinforces the parallel evolution in awareness and behaviours the individual guide encourages in your emerging leaders.

Part 3
Refinding Faith
Transforming the Leadership Team

KEY POINTS

▸ The critical awareness that Mission Control lost and rediscovered in our management roles is that everything we do contributes to the organization's success or failure, not just in our individual jobs but also in the entire organization's core purpose.

▷ Finding our way back from 'no ripples in the pond' to the real-time morality in our management roles first required us to relearn this awareness.

▷ Taking the next step to evolve a team to high-trust leadership hinges on embracing the real-time morality, not just at the working level but also as managers.

▷ Rather than allowing the management cloud to dilute the connection to their core purpose, the real-time morality empowers a leadership team to preserve the connection, just as it does for the team in the Mission Control Room.

▸ Embracing the deliberate morality and evolving to a high-trust leadership culture and management practices is repeatable in any leadership team.

▷ The morality and awareness empower a leadership team to have the uncomfortable discussions, make critical judgements and choose to be as deliberate in all facets of management as in any critical decision-making.

▷ With this focus, the team can then explore ideas to increase trust, understand and articulate core values and purpose, and align as a leadership team.

▷ The resulting leadership culture entails high-trust and fully transparent engagement among the senior leaders in the same way that successful spaceflight requires high trust and full transparency from the flight control team.

▶ By leveraging our experience as a head start in awareness, and by using the team guide as a roadmap, this evolution can be done more deliberately and in less time than we took in our discovery.

▷ The four philosophical evolutions to lead a team through are summarized by:

1. Align the team to core values.
2. Incorporate transparency and high-trust behaviours as team norms.
3. Anchor all management strategies and decisions on deliberations in this transparent, aligned and engaged team.
4. Reinforce an ongoing integrity focused on deliberate alignment and high-trust behaviours.

▷ The team evolution and individual development rely on the book-club experiences and foundational ideas that were key to Mission Control's evolution and are laid out exquisitely in the following five books:

1. *What Got You Here Won't Get You There*, Marshall Goldsmith.
2. *The Speed of Trust,* Stephen M.R. Covey.
3. *Leading Change*, John P. Kotter.
4. *Good to Great,* Jim Collins.
5. *Built to Last*, Jim Collins and Jerry I. Porras.

▷ The key is open and thorough dialogue to internalize the ideas and observations they help us make, followed by deliberate and related management-practice changes with the specific intent of maximizing transparency amongst the leadership team.

▷ Transparency leads to increased willingness to critically evaluate current practices, explore more ideas and develop an openness to change aligned with values and a common core purpose.

▸ Deliberately developing the managers who are at the level from which you will promote into your Senior Leadership Team primes the pump with candidates who have already evolved their individual-leadership styles and are ready to further the Senior Leadership Team's efforts.

▷ In developing emerging leaders, we apply the book-review process that emphasizes peer and supervisor discussion of the key points, focusing on key takeaways, personal strengths and weaknesses, and areas for both the new manager and the team to work on going forward.

▷ The goal is to normalize: regular discussions focused on transparency, values alignment and engagement; the willingness to judge individual and team behaviours; the stewardship of these leadership values on behalf of the organization's core purpose.

▷ It is this self-awareness and mindfulness that also reinforces a leader's high-trust management style and enhances organizational performance.

Bon Voyage!

Rocket science is easy. People is hard.

J. Milt Heflin, Chief of the Flight Director Office, 2003

In any journey, without a first step, we go nowhere. The critical first step is deciding to either set out or settle for where we are. In this leadership journey, that means we choose either to pass judgement on our current leadership environment, or take the path of least resistance, 'dance with the one that brung us', no ripples in the pond etc. The specific next steps we could take are much less important until we learn an awareness that we have areas in which we can improve, and bridge from that awareness to a willingness to explore changes that can get us there.

We've seen where not having that awareness can lead, even in established organizations with long records of admirable success. The management practices MOD adopted through 'diplomacy over clarity' led us away from the values – the real-time morality – that we still saw as so essential in our most critical work, costing us not only the trust and support of our stakeholders and customers but also an increased risk of failing in our core purpose. That same lack of awareness at even higher levels led NASA to similar management practices that contributed directly to each catastrophe that cost us our astronauts' lives.

MOD's experience shows that learning and accepting that awareness – putting the ripples back into the pond – can be a difficult, emotional and even traumatic choice for individual leaders and a team. For us, it took more than the deteriorating trust from our stakeholders and customers and near misses in our work. We didn't take that first step until we were confronted with other teams who excelled in work that we considered to be core capabilities. When we saw how far behind we were in some of our most important work, we were compelled to reconcile that contrast and do something about it.

So began our leadership journey. We did not know where it would lead, only that we had to go. With that awareness pushing us along, we were empowered to pass judgement and apply new ideas, which led us to real management behaviour and practice changes. The high-trust leadership environment we put in place through this journey brought us the same highly reliable decision-making and strong performance in our management roles that are critical in the Mission Control Room.

Rather than waiting for your own brush with failure and searching for ideal benchmarking opportunities, use our experience and our journey to get your team's attention. Consider this not just a case study but a benchmarking report from before and after the leadership evolution in Mission Control's management ranks. Start the awareness that can help you break out of the paradox of success and the management cloud.

Then set out as a team on your journey.

Leverage the ideas that delivered first a sea change and then a tidal wave of deliberate leadership and management practices in Mission Control. Your core purpose, strengths, weaknesses and risks may not be the same as ours, nor will your specific management-practice changes. But the real destination isn't those specifics – it is to enable a leadership environment that inherently strengthens your performance in achieving your core purpose. The real-time morality – technical truth, integrity and courage – enables this leadership environment while managing our most critical risks.

Discover as a team how to enable this leadership environment in your management roles through deliberate transparency, values alignment and engagement, just as we did. Find and make peace with your individual and team awareness. Embrace the morality to reinforce mindfulness. Pass judgement. Take action through transparency, values alignment and engagement with your team.

And then keep going.

This journey never ends, because the management cloud is always there to lure us into behaviours that erode the leadership environment despite our best intentions. The opportunities for mistakes, from small to catastrophic, are also always with us.

Keep stoking that awareness and the willingness to pass judgement. Seek new ideas and perspectives, apply them to your behaviours and management practices through open team discussion, and keep reinforcing this as the management norm at all levels of your organization. It can then take root as a culture at all levels, not just in the Mission Control Room.

Lead deliberately and lead well.

It ain't rocket science. It's much more difficult and more important than that.

Notes and References

Praise for *Out of This World*

1. Blanchard, K. & Johnson, S. (2015) *The New One Minute Manager*, William Morrow
2. Blanchard, K. (2009) *Leading at a Higher Level: Blanchard on Leadership and Creating High Performing Organizations*, Prentice Hall

PART 1

CHAPTER 1

3. 'last line of defense': Kranz, G. (2009) *Failure Is Not an Option: Mission Control from Mercury to Apollo 13 and Beyond*, p.131, Simon & Schuster
4. Chris Kraft quote from his speech at MIT: Hoffman, J. (2005) *16.885J/ ESD.35J Aircraft Systems Engineering, Fall 2005,* Massachusetts Institute of Technology: MIT OpenCourseWare, http//www.ocw.mit.edu
5. 'discipline': Kranz, G. (2009) *Failure Is Not an Option: Mission Control from Mercury to Apollo 13 and Beyond*, pp.126–31, Simon & Schuster
6. 'Spaceflight will never tolerate': (2013) 'Through a New Lens: Apollo, Challenger, and Columbia through the Lens of NASA's Safety Culture Five-Factor Model', *NASA Safety Center System Failure Case Study*, Volume 7, Issue 3, p.5
7. Pete Frank quote: Frank, P. (1980) 'The Foundations of Flight Control', NASA JSC memorandum CF-80-119-2j
8. Jeff Hanley quote: Hanley, J. (2005) 'The Foundations of Mission Operations', NASA JSC memorandum DA8-05-060

9. George C. Marshall quote: Marshall, G.C. (1950) *The Armed Forces Officer,* p.58, United States Government Printing Office

10. Stone Tablets revised from: Stone, B.R. (1988) 'The Flight Directors' Stone Tablets of Flight Controller Operations', informal NASA MOD

11. Excerpt: Bourque, D. (1981) 'The Autonomous Flight Controller', informal NASA MOD

12. Revised cue card from: 'Links of the Error Chain', informal NASA MOD, ~1990

CHAPTER 3

13. Albert Einstein quote: Calaprice, A. (ed) (2010) *The Ultimate Quotable Einstein,* pp.384–85, Princeton University Press

14. 'Covey sums up trust': Covey, S.M.R. (2006) *The Speed of Trust: The One Thing That Changes Everything,* p.30, Free Press

15. Chris Kraft quote from his speech at MIT: Hoffman, *16.885J/ESD.35J Aircraft Systems Engineering, Fall 2005*

16. Shuttle flight rule: *Space Shuttle Operational Flight Rules – Volume A – All Flights: NSTS 1282 – PCN-8* (2007) pp.2–252, NASA Johnson Space Center

17. 'This determination': Kraft, C. (2001) *Flight: My Life in Mission Control,* p.272, Dutton Books

PART 2

18. Emerson quote: Emerson, R.W. (1876) *Letters and Social Aims,* p.86, James R. Osgood and Company

CHAPTER 4

19. Pete Frank quote: Frank, P. (1980) 'The Foundations of Flight Control', NASA JSC memorandum CF-80-119-2j

20. Space Program Costs: Lafleur, C. (8 March 2010) 'Costs of US Piloted Programs', *The Space Review*

21. 'Project Apollo: A Retrospective Analysis', NASA History Home, April 2014, http//www.history.nasa.gov/Apollomon/Apollo.html

22. MOD organization chart: Derived from NASA JSC MOD, August 2011

CHAPTER 5

23. 'We were a ruthless bunch': Kraft, C. (2001) *Flight: My Life in Mission Control,* p.117, Dutton Books

24. 'Aligning our Behaviors': Flynt, G.A. (2005) 'Achieving Mission Success – Aligning our Behaviors with Our Values', pp.1–2 informal NASA MOD

25. 'People in MOD, 2006': Woods, E. (2010) 'MOD Financial Summary', informal NASA MOD

26. 'We were too gung-ho': Kranz, G. (2009) *Failure Is Not an Option: Mission Control from Mercury to Apollo 13 and Beyond,* p.204, Simon & Schuster

27. 'This was a meeting where': Presidential Commission on the Space Shuttle Challenger Accident (1986) p.94, *Report to the President,* Government Printing Office

28. Challenger details: Presidential Commission on the Space Shuttle Challenger Accident (1986) *Report to the President,* pp.19–21, 129–31 Government Printing Office

29. 'The Commission concluded': Presidential Commission on the Space Shuttle Challenger Accident (1986) *Report to the President,* p.104, Government Printing Office

30. 'Within NASA, the cultural impediments': Columbia Accident Investigation Board (2003) *Columbia Accident Investigation Board Report, Volume 1,* p.208, Government Printing Office

31. 'NASA is a federal agency': (2003) *Columbia Accident Investigation Board Report, Volume 1,* p.6, Government Printing Office

32. Jim Collins quote: Collins, J. (2009) *How The Mighty Fall: And Why Some Companies Never Give In,* p.8, Random House Business

PART 3

CHAPTER 7

33. John Kotter quote: Kotter, J.P. (1996) *Leading Change*, p.vii (2012 Preface), Harvard Business Review Press

34. 'Eight-Stage Process': Kotter, J.P. (1996) *Leading Change*, pp.33–158, Harvard Business Review Press

35. 'Cultural Behaviors': Covey, S.M.R. (2006) *The Speed of Trust: The One Thing That Changes Everything*, p.237, Free Press

36. 'The moment you feel' and three other quotes from Jim Collins: Collins, J. (2001) *Good to Great*, pp.56, 58, 74, 86, HarperCollins

37. 'Distinguish core values' and 11 other quotes from Jim Collins: Collins, J. & Porras, J.I. (2004) *Built to Last: Successful Habits of Visionary Companies*, pp.xix, 166, 167, 228, 137, 139, 228, 173, 121, 236, 167, 121, HarperBusiness

CHAPTER 8

38. Marshall Goldsmith quote: Goldsmith, M. (2007) *What Got You Here Won't Get You There: How Successful People Become Even More Successful*, p.21, Hyperion

39. 'The paradox of success' and four more quotes from Marshall Goldsmith: Goldsmith, M. (2007) *What Got You Here Won't Get You There: How Successful People Become Even More Successful*, pp.17–24, 21, 21, 24, 112, Hyperion

CHAPTER 9

40. Jim Collins quote: Collins, J. & Hansen, M.T. (2011) *Great by Choice: Uncertainty, Chaos, and Luck – Why Some Thrive Despite Them All*, p.21, HarperBusiness

41. 'People in MOD, 2009': Woods, E. (2010) 'MOD Financial Summary', informal NASA MOD

42. 'Yes, leadership' and 'The Stockdale Paradox': Collins, J. (2001) *Good to Great*, pp.74, 86, HarperCollins

43. 'People in MOD, 2011–2015': Woods, E. (2014) 'MOD Financial Summary', informal NASA MOD

CHAPTER 10

44. Stephen M.R. Covey quote: Covey, S.M.R. (2006) *The Speed of Trust: The One Thing That Changes Everything*, p.298, Free Press

LEARN FROM A VETERAN MISSION CONTROL AND EXECUTIVE LEADER *IN PERSON*!

This book is just one facet of Paul's 'leadership evangelism'. He is also a sought after speaker, and his Mission Control leadership insights have already inspired leaders in audiences which include Fortune 100 companies, professional conferences, universities, executive development programs and more.

Paul leverages his critical leadership experience and offers insights that can be applied in any business – whether the goal is to strengthen working level decision making and leadership; actively steward leadership values; or improve your team's performance in strategic decision-making and leading change.

His keynotes delve into topics including: how so many top performers turn out to be lousy leaders, deliberate cultural stewardship, critical incident preparedness, high-performing teams and more.

Want to go even deeper? Paul leads a one-day workshop that will describe Mission Control's revered brand of leadership and how to make the leadership culture your own. This is your opportunity to learn it from someone who has led at all levels from the Mission Control Room to the boardroom and revolutionized the leadership culture.

The workshop will take you from the ground up in the values that are most critical to highly reliable team performance. You will then apply the leadership culture from the top down as deliberate organizational values and management practices, and learn to reinforce the performance where it matters most every day.

Failure is always an option, and so is choosing to lead your team into a leadership environment that helps them avoid catastrophe and pull off miracles. Let Paul bring you insights to help take them there.

Get started with Paul Sean Hill and Atlas Executive Consulting at:

pshill@AtlasExec.com   www.AtlasExec.com

Would you like your people to read this book?

If you would like to discuss how you could bring these ideas to your team, we would love to hear from you. Our titles are available at competitive discounts when purchased in bulk. Bespoke editions featuring corporate logos, customised covers or letters from company directors in the front matter can also be created in line with your special requirements.

We work closely with leading experts and organisations to bring forward-thinking ideas to a global audience. Our books are designed to help you be more successful in work and life.

For further information, or to request a catalogue, please contact:
business@johnmurrays.co.uk

Nicholas Brealey Publishing is an imprint of
John Murray Press.

Thank you very much for reading our story and learning to apply our experiences in your business. Lead deliberately and lead well!

If you like this book, please share your experience through a review at Amazon.co.uk and Amazon.com.